Effective Climate Communication

Anastasia Denisova

Effective Climate Communication

Turning Eco-Anxiety into Eco-Action

Anastasia Denisova
University of Westminster
London, UK

ISBN 978-3-031-67339-9   ISBN 978-3-031-67340-5 (eBook)
https://doi.org/10.1007/978-3-031-67340-5

© Springer Nature Switzerland AG 2025

This work is subject to copyright. All rights are solely and exclusively licensed by the Publisher, whether the whole or part of the material is concerned, specifically the rights of translation, reprinting, reuse of illustrations, recitation, broadcasting, reproduction on microfilms or in any other physical way, and transmission or information storage and retrieval, electronic adaptation, computer software, or by similar or dissimilar methodology now known or hereafter developed.
The use of general descriptive names, registered names, trademarks, service marks, etc. in this publication does not imply, even in the absence of a specific statement, that such names are exempt from the relevant protective laws and regulations and therefore free for general use.
The publisher, the authors and the editors are safe to assume that the advice and information in this book are believed to be true and accurate at the date of publication. Neither the publisher nor the authors or the editors give a warranty, expressed or implied, with respect to the material contained herein or for any errors or omissions that may have been made. The publisher remains neutral with regard to jurisdictional claims in published maps and institutional affiliations.

This Palgrave Macmillan imprint is published by the registered company Springer Nature Switzerland AG
The registered company address is: Gewerbestrasse 11, 6330 Cham, Switzerland

If disposing of this product, please recycle the paper.

To George, Paolo, Tatiana, Vladimir and Vadim

Acknowledgments

This book became possible thanks to the work of exceptional journalists—those who present climate change stories in an engaging, moving, people-focussed yet solution-oriented way. Jeff Goodell's long feature about mosquitos and humans in Rolling Stone, as well as ProPublica and The New York Times Magazine's 'Where will everyone go?' awed and shocked me the first time I read them a few years ago. As good journalism does, they gave me a glimpse into the realities of the others, in the most empathetic and alarming way. They called me to action, and they showed that the case is not lost—a lot can still be done.

I am most grateful to the two reviewers who kindly encouraged me to go ahead with this book at the proposal stage—and I am indebted to the wisdom, kindness, and most insightful analytical comments of the reviewer of the full manuscript.

So many wonderful colleagues helped me along the way—as inspiration, mentors, role models. In no particular order, I would like to thank Prof Anna Feigenbaum, Prof Roza Tsagarousianou, Prof Graham Meikle, Dr Caroline Ruddell, Dr Anastasia Kavada, Dr Andrea Medrado, and Dr Nathasha Fernando. My colleagues at the University of Westminster have always been a great source of inspiration.

I have been hugely inspired by the work of Prof Katharine Hayhoe from Texas Tech University, a respected scientist who advocates for constructive, unrelenting climate communication with a heart.

Big thanks goes to Palgrave Macmillan, to my thoughtful editor Lauriane Piette and the super-organised Alice Carter and Eliana Rangel.

Big thanks to my family for buying me coffees whenever I felt in doubt.

Thanks be to God for guidance and giving us a chance to do better.

June 2024   Anastasia Denisova

Contents

1 Introduction 1
   Bibliography 10
2 Ten News Values for Climate Communication: From
  'Crisisation' to Attribution, Emotional Offsets,
  Pragmatic Instructions and Punchy Storytelling 13
   *No Pain No Gain? Why Traditional News Values Are
   Outdated for Climate Coverage* 15
   *Cognitive and Emotional Undercurrents of News Fatigue
   and Disempowering Storytelling* 17
   *Climate Change and News Values—More Agency, Good
   News and Mid-level Solutions* 20
   How High Earners and Social Influencers Can Be Helpful ... 46
   Mid-level Activity and Solutions 48
   Empowerment 50
   Bibliography 52

3 Global South and Global North: Discrepancies
   in Climate Coverage   61
   *Global South—How Is Climate Covered in the Countries
   that are Hit the Most. Issues with Delocalisation, Resources
   and Literacy*   61
   *Foreign Aid, and How the Media Shape the Agenda
   in Inter-country Giving*   68
   *Global North—Progressive Advancements, Precarious
   Consensus, Political Polarisation and Powerful Lobby*   70
   *The Most Prominent Five Frames of Climate Coverage
   in the Global North*   76
   Bibliography   79

4 The Many Faces of Greenwashing   83
   *How the Coffee Takes the Worry Away, or a Tale
   of Coffee-Washing*   83
   *Doubt-Mongering, Diluted Terminology, and the Carbon
   Footprint Calculator Paradox*   87
   Bibliography   93

5 The 'Ignorance as a Choice' Paradox, and the Role
   of Depleted Resources in the Responses to Climate
   Messages   97
   Is Ignorance a Route to Happy Life?   97
   *Pivot to Emotions—Why a Tonic of Fear and Hope Is
   Desired for Climate Action*   99
   *'The Ignorance Explosion' Paradox and Heuristic Bias
   in the Context of Accelerated Societies*   103
   Eureka, or How Human Psyche Chooses the Quick Answer   106
   *Why People Deny Climate Change, or the 'Dragons
   of Inaction'*   107
   Bibliography   110

6 From Emotions to Determination: The Communication
   Tools for Free Riders and 'Conditional Cooperators'   113
   Bringing the Hearts to the Table, Not Just the Minds   116
   The Gift of Climate Action   118
   *In for a Ride? The Moderation Techniques for Free Riders
   and 'Conditional Cooperators'*   119

	Public Opinion Surveys	120
	Social Proof as the Foundation of Social Norm Construction	124
	Social Proof Contextualisation and Challenges Within Social Media	131
	The Structuring of Human Experience—The Dichotomy of Individual and Collective Sensemaking	133
	The Fascination with Positivity in Social Identity, and the Repertoire of Pro-climate Identities	137
	Identity Threat, and How the Media Shape a Great Deal of It	142
	Visual Power Struggles, or Why Greenpeace Is Loading Glaciers and Bears with Meaning	145
	Bibliography	149
7	**Climate Optimism or Climate Pessimism? Self-Efficacy Boosters and Storytelling for Change**	157
	Get Your Jab, or How Just One Tweak of Communication Makes People Act	157
	Eco-anxiety and Turning Strong Emotions in the Currency of Change	159
	The Route to Eco-empowerment—Self-Efficacy	161
	Apocalyptic Storytelling: Why It Awes, but in the Wrong Way	168
	Is Climate Optimism an Empowering Strategy or a Route to Complacency?	173
	What Anti-drunk Driving Campaigns Can Teach Climate Communicators?	176
	Shaming Me, Shaming You?	180
	The Hero Narratives, or How Humans Like to Crown Their Idols and Then See Them Fall	182
	The Powerful Acts of Storytelling—A Tool for Intervention and Climate Empowerment	188
	Bibliography	190
8	**Conclusion**	197
Index		201

About the Author

Dr. Anastasia Denisova is a Senior Lecturer in Journalism at the Communication and Media Research Institute, University of Westminster. She specialises in viral cultures, Internet memes and climate change communication. She published a book *Internet Memes and Society* in 2019. In 2021, Dr. Denisova undertook an extensive review of fashion coverage in the media and released a policy brief *Fashion Media and Sustainability*. Anastasia has published widely in top academic journals, including *Social Media+Society; Media, Culture and Society; Journalism;* etc. Dr. Denisova is a Senior Fellow of Higher Education Academy and a Board member of *Westminster Papers in Communication and Culture*.

CHAPTER 1

Introduction

We live in the times of media abundance. From traditional journalism organisations to commercial social media platforms, and public service and community initiatives in between, an average person is inundated with news, messages, notifications, in other words—the stimuli that affect cognitive and psychological systems. They require a reaction. Therapists compare this bombardment of information with working as 999 operators but without professional training or any days off allowed.

Climate change, as a long-term issue, competes for attention with multiple other crises—short term, long term, singular or repeating—that generate headlines and notifications. In 2019, the British broadsheet and website The Guardian proposed the terms 'climate crisis' and 'climate emergency' or 'breakdown' as the preferred vocabulary to discuss the anthropogenic changes in nature and habitats. The editor in chief criticised the expression 'climate change' as 'rather passive and gentle' (Carrington, 2019). Ever since, the organisation adopted a more forceful narrative around climate change, emphasising 'crisis' and documenting its many manifestations. As commendable as this turn of narrative was in its undertaking to draw public attention to the severity of the climate change, it cannot remain a sustainable *modus operandi* for the media and citizens alike. While the urgency of the climate change has been pronounced by many major media outlets in the West and beyond, this book argues for the variation in frames and choice of stories to **maintain** the attention of the exhausted readership.

This 'crisisation' is nothing new. Newspapers do not talk calmly—they scream about the events and developments in need of attention. They do so to ring the alarm to help those who might be affected, but also to compete for the eyeballs and clicks of the wandering audiences. Professor Roza Tsagarousianou criticises the pervasiveness of crisis framing—in her case, it refers to the media coverage of migration and asylum seekers. Confining people's movement to the frame of 'crisis' (Ramsay, 2020; Tsagarousianou, 2023) keeps the individuals locked in the migrant identity, denying them the chance to blend in with the population and exceptionalising them, 'Othering' them from non-displaced citizens. Applying the crisis frame to climate change coverage bears similar overtones—while it pursues the aims of raising the awareness of the public and politicians to the immensity of the issues at hand, it separates the realm of climate change from people's everyday existence. The immensity of climate 'crisis' makes individuals feel small and irrelevant; it refers to the scope of events happening on a global scale, literally way above our heads, in the atmosphere and all its elements—the issue of 'Othering' of climate change becomes ever more real if the media persist presenting it as a realm separated from daily living and comprising distressing headlines that a person cannot possibly affect.

We live in the chronotope of many emerging—but not solved—crises. Applying Mikhail Bakhtin's (1986) notion of 'chronotope'—the conflation of spatial and temporal dimensions in storytelling—allows to see how the media narratives on climate change depict the state of crisis as an intrinsic way of being. Readers are experiencing time deeply compressed with space—at any given point during the day something awful is happening somewhere, as the news notifications attest, and the rhythm of living seems to be devoid of escaping this. The permanency of climate change, in long term, and its fluctuating timeline of developments, in short term, present a dual challenge to the chronotope of crisisation. The crisisation mode embraces the acute aspects of climate change, but Others the chronic, overarching issues that need to be confronted. This contentious state of events is further complicated by information uncertainty—another term explored further in the book—which explains the vicious cycle of readers trying to mitigate the anxiety created by the crisisation through more consumption of the crisis-ridden media. This often leads to problematic outcomes.

The crisisation of climate change cannot be the permanent frame of seeing and assessing the climate-related events. The crisisation of climate narratives offers the overstimulated citizens of the overwhelming mediascape the temptation to 'Other' this massive global happening, disconnect from the vastness of its causes and consequences. The 'present shock' (Rushkoff, 2014) that many people experience every day—the vividness and detailed richness of the terrible and uplifting events we see through the media lens—can result in a permanent state of anxiety (Dhir et al., 2018; Keles et al., 2020). Anxiety is a draining psychological condition—a healthy body and mind is seeking ways to reduce its effects, likely through various coping mechanisms, as well as disconnection from the 'crises'.

Advancing from the unsustainability of crisis narrative as a starting point, this monograph delves into many layers of communication on climate change.

Chapter 2 analyses how the media still rely on the set of traditional news values that need an urgent review—climate change is a long-term issue that the audiences and media professionals must adapt to; it is not a fleeting crisis that can fit the classic news span of five days or so and give space to a new trending story. My chapter on ten news values adapted for climate coverage proposes more focus on the frequency of extreme weather events and attribution, establishing events in a pattern that would explain the trajectory of climate change to the audiences, not shock them with every new disaster as if it came out of the blue, one disconnected from the other. The media can look not only at the crises unfolding, but at how people are rising against the adversity. Bad events need to be reported—but with clearer links to how climate change amplified their probability, and what actions are being taken to prevent new events of the sort and how to help the communities adapt to the next event that is likely to come. This approach brings in the advantages of solutions journalism and constructive journalism, while aiming to keep the audience engaged with the issues, not withdrawn due to the rising anxiety or a sense of being overwhelmed.

The climate-adjusted news values position a magnifying glass over the stories of innovation and empowerment, as well as investigations of greenwashing and carbon offset claims. The media are urged to shed more light on the collectives that bring change in climate mitigation and adaptation—not just individual activists, but industries and workplaces that make a pivot to going green. The adapted news values need to refer to awe and prioritise the powerful acts of storytelling that refresh people's understanding of the humongous work that is being done to make a dent in the problem, and the stories that keep audiences emotionally connected to climate change.

Chapter 3 looks at the patterns of climate coverage across the globe. It celebrates the fact that the polarisation of opinion on the origins of climate change—whether climate change owes to human activity or not—is an issue of the past in many parts of the world, including many countries in Europe, the UK, as well as China, Nigeria, Kenya, Ghana, South Africa, several South American countries. The BBC (a public service media company in the UK) only stopped platforming climate deniers in 2018, thus demonstrating that even the most trusted and prestigious media institutions were slow in setting the consensus on the anthropogenic origins of climate change. The news values of conflict, proximity and scale have dominated the climate coverage for decades—and grew outdated for the crisis we face today. The televised conflict between 'pro-climate change' scientists and climate deniers is, luckily, becoming a relic of the past—it lasted because it fitted the 'conflict' value and the misguided sense of journalistic objectivity. Currently, in many parts of the world, the need to act upon climate crisis is pronounced on the highest levels of governance and is reiterated through influential media organisations and businesses.

Yet a different kind of polarisation—designating climate crisis as a political issue and arguing over the degree of sacrifices that countries and businesses needs to commit to in order to minimise the damage—is manifesting greatly in the media coverage in the US and Australia, Canada, to some extent in the UK (at the time of writing in 2024), among other examples. This is due to the historically ideological nature of climate conversations in some countries. For instance, in the US, the 2000 Democrat Presidential candidate Al Gore mentioned pro-climate policy and scientific discussions in campaigning, and then went on to win the Nobel Peace Prize for his climate change documentary in 2007; while in the likes of Australia and Canada, the economies rely on fossil fuels thus impeding grander action to reduce those.

Another, non-ideological, issue that permeates global coverage of climate crisis and slows down citizen action—to a much larger extent in the 'Global South' than 'Global North'—is the lack of climate literacy and newsroom resources to cover the biggest story of our times. The attribution confidence—linking extreme weather events to climate change—demands an avid level of knowledge of the roots and effects of climate data, skills to work with the historical data and compare them to the present. The newsrooms are limited in time and resources, which prevents them from delving into the issue, from reporting from the communities and industries affected on the ground, and engage with climate scientists to obtain expert commentary on their perspective—all these practical hurdles have a profound effect on the quality and consistency of climate coverage in the unforgivingly fast-paced media environment.

The lack of human stories and reporting of the impact of climate change on the people and livelihoods in the countries affected the most was among the most striking findings of this book. The 'delocalisation' of climate change prevents the communities from adapting to the issue and from having the right tools to advocate for their survival. There is a visible distortion of the global climate agenda—developed countries originate heaps of climate stories, which are then echoed by the coverage in the developing countries; this is particularly evident during major Western climate summits like COP and the releases of the scientific reports such as that of the IPCC (Intergovernmental Panel on Climate Change). This tendency establishes Global North as the agenda-setting hub while simultaneously diminishing the agency in the Global South.

Greenwashing, doubt-mongering (Oreskes & Conway, 2010) and carbon offsets communication are right in the middle of this tome, in Chapter 4. The chapter on greenwashing concentrates the complexity and the ambition of the climate change action—but for all the wrong reasons. There is no doubt that the communication of a fundamentally complex global issue needs to be boiled down to simpler, understandable, practical terms and actions. Yet this is exactly what makes greenwashing—the custom of businesses and countries to claim big changes while continuing their planet-heating practices—so appealing and viral. It manifests in the likes of the carbon footprint calculator and carbon credits. The first tool, developed by a fossil fuel company over a decade ago, diverts attention from the big challenge to phase out fossil fuels and blames individual consumers instead. It suggests individuals should look at themselves first. The second initiative—that started as a market tool to curb the emissions—has resulted in building a new multi-million industry of the companies that promise a sort of Medieval church indulgences. An example of this is the airlines that can keep operating normally, as long as they buy some 'offsets' from a designated company that promises to plant more trees or invest in green public transport in remote lands. The dubious claims and the diversion of attention in both cases call for more literacy and critical thinking from media practitioners, policymakers and audiences.

Chapter 5 gets intrigued by the power and weakness of knowledge, and the pride of ignorance that the accelerated societies have brought upon many citizens. It follows the Nobel Prize winner in Economics Daniel Kahneman in evaluating the limited cognitive storage for facts and emotions that humans possess. It then scrutinises the formidable influences exerted by egocentric impulses and altruistic tendencies on individuals' propensity to either engage with or abstain from addressing climate change. It dedicates significant attention to the dynamics involving free-riders and 'conditional cooperators', colloquially referred to as humans. The majority of us oscillate between these behavioural dispositions of charitable and uncharitable behaviours contingent upon contextual factors. This chapter discusses why more knowledge is not an answer to the lack of interest in climate stories—but the alignment with individual values, principles and interests could help. Katharine Hayhoe (2021), a prominent Canadian American climate scientist, encourages people to talk about climate to anyone who would listen. The research in this book demonstrates that the normalisation of climate conversations is linked to the increasing public awareness of the 'norm'. Both climate change worry and climate change action are the new components of life in the twenty-first century—they are the acts that cannot be left to the politicians and elites alone. The media have the power in constructing the norm through narratives on climate change-related events; they affect our private conversations and public debates. It is essential to prevent the 'Othering' of the climate change in public consciousness and ensure that the 'pro-climate action mindset' is presented as the default one.

Chapter 6 sheds light on the multifaceted accomplishments of social psychology in relation to the media framing of climate change. It analyses theories such as 'social proof', free riders, conditional cooperation, social identity theory, social comparison and social norm construction, and applies them to underscore the media's function in establishing societal norms and mirroring society's collective identity. The media hold a powerful role in educating, informing the citizens, but also consolidating the social norm of climate change awareness and interest, and taking an active position in mitigating it.

The next step, as this book proposes in Chapter 7, is for the media and public institutions to provide clear and efficacy-boosting promptings on *how* to act upon this emerging consensus. Since the seminal studies of Albert Bandura on self-efficacy building, it has never been a better moment to apply the 'theory of small steps', as one may conveniently nickname his approach, to climate actions. If you only have time to read one chapter, read Chapter 7. It warns against doomism and apocalyptic storytelling and provides important reflections on the archetypes and powerful acts of storytelling.

Chapter 7 concludes with the forward-looking thoughts on the hero storytelling arch, and why individual heroes are helpful as the ignition tool but need to be strengthened by collective action narratives for a long-lasting effect. The chapter on forward-looking media frames utilises the studies on drunk-driving campaigns that demonstrate the paradox—although we all know what is 'the right thing' to do, i.e. not drink and drive, there is a 'bug' in the human nature. The same rightful message has to be repeated again and again—in novel and creative formats every time—to keep it fresh and relevant. Climate change action needs a similar creative and persistent approach—it is an open call for plenty of creativity and commitment from the media and policymaker minds in the years to come.

Chapter 8, Conclusion, brings many discussions and enquiries of this book together. In a nutshell, this monograph makes eight forward-thinking contributions to media studies:

1. It makes a strong science-based argument **to retreat from climate doomism** as it alienates the audience. It proposes a feasible range of strategies and techniques that help media professionals to adapt to the long-term problem that is climate change.
2. It provides a timely review of journalism news values and modifies them for climate coverage. This results in **10 highly applicable climate coverage news values** that can keep the audience engaged with and empowered by climate storytelling.
3. It develops a new theoretical framework for media studies in applying the **social proof** framework to the mediated construction of common sense. Thus, the monograph conceptualises a progressive, anxiety-reducing heuristic in battling climate disengagement, nurturing the sense of community and managing information overload in the audience.
4. It evaluates the many issues with and layers of greenwashing in the media; it takes a step further to propose the concept of **'greenwishing'** as the designation of superficial lifestyle advice that

gives a false sense of action but becomes an excuse for climate delayism.
5. This research advocates for the implementation of **tailored instructions and efficacy boosters** across media coverage, targeting specific groups, communities, backgrounds, ages and genders, among other characteristics. Individuals need to become aware of their power to address the climate crisis on individual, workplace and community levels, and to connect this awareness to their understanding of the developments on national and international levels.
6. This research encourages the implementation of targeted, meticulously researched campaigns that are cautious about the use of **shame- and guilt-inducing storytelling** frames. The excessive use of shame and guilt in climate campaigning can backlash and result in lower engagement and efficacy among audience members.
7. The book argues for the increased use of **the powerful acts of storytelling**, such as documentaries or 'view from the moon' stories that rely on strong human-interest narratives, catchy multimedia execution and emotional appeals, along with the clear embedded facts. Documentaries are known to generate global interest from audiences and may serve as an effective communicative intervention to inform, engage and motivate action. Long-form features and podcasts hold similar promise.
8. While the powerful classic storytelling arch of the Hero's Journey remains a recurrence in media coverage on climate issues, this book cautions **against the overuse of the 'lone hero' storytelling frame and recommends focussing on community examples and collectives** that represent diverse classes, ages, genders, incomes and locations. These examples are more relatable to a broader range of people who care about climate change but do not see themselves represented in the media in positions of agency and impact. The lone hero narrative—or a Hero's archetype—is attractive in the short term for directing the attention and sense of empowerment around the issue. However, there are shortcomings in the crowning of individuals that represent the whole issue, including the inherent human desire to see heroes fall and challenge the idols.

Through a critical theoretical journey, multi-disciplinary analysis and scrutiny of case studies, this book contributes the pragmatic and efficient modes of engaging the public with media communication on climate. It follows the definition of engagement as personal connection on cognitive, emotional and behavioural levels (Lorenzoni et al., 2007). These characteristics are complex and not connected in a linear fashion (Whitmarsh & O'Neill, 2011)—they provide the grounds to inspect the uneven, dynamic interplay of knowledge, intentions, feelings and actions. The range of actions that prove that individuals and communities are acutely aware of the damage of the climate change can vary from consumer decisions to mobilisations and political activism.

The term 'empowerment' is defined in this book as the cognitive-emotional response of interest and eagerness to act towards climate change, with the pronounced efficacy felt by the members of the public. What this book is aiming to achieve is empowering a society of climate-conscious citizens that embed climate adjustments and action in their daily lives—and demand the same from their politicians and elites.

The choice of terminology regarding 'climate change' or 'climate crisis' in this book is not random. For the reason of approaching climate change as a process, not a one-off issue to fix, this book suggests avoiding the terms 'climate crisis' and 'climate emergency' in much of journalism coverage for the public. It is unsustainable for a human system to stay in a constant state of alert or 'emergency'—it leads to the burnout and withdrawal from the issues. This is why 'climate change' and 'global heating' are more sustainable terms for the media practitioners to use to discuss the new way of living. There may be an occasional nod to 'climate crisis' for the sake of finding synonyms to the main terminologies, but it should not be the major frame to cover the issue. We will be living in the shadow of this colossal global process for the decades, if not centuries, to come, and adapting storytelling to this new mentality is a real task for sustainable journalism.

The media cannot solve all the problems. Yet, despite the rise of social media platforms and various critical issues that come with it, including the non-transparency of algorithms, commercial business models and growing disengagement of young people with the media, my research for this book has demonstrated that the professional media still set the agenda. They still matter. And now, more than ever, when climate issue is in the headlines and in the minds, it is crucial not to lose the momentum and keep people *actively* engaged with the matter. As Roberto Cialdini showed in his multiple experiments over the decades, it is a small donation, small involvement, small sacrifice and small decision to read more about the political candidate's climate promises, about the carbon footprint of big companies and impact of one's lifestyle choices—that can build the commitment to the cause, encourage further interest and expansive activity.

The book's limitations lie in a reduced analysis of social media—therefore, further research is encouraged to examine the fluctuating patterns of climate communication on social networks, and a thorough scrutiny of the pervasive yet adaptive algorithms that underpin that, as well the artificial intelligence-powered communication tools that will play a larger role in gatekeeping in the years to come.

The strength of this book lies in the scientifically backed belief in the power of storytelling and quality journalism, the compassionate and pragmatic assessment of the cognitive and psychological capacities and limitations of humankind in the processing and assessment of information, and the forward-looking amplification of the narratives, frames and communication techniques that can engage and empower the audiences in the age of climate change.

Bibliography

Bakhtin, M. (1986). *Speech genres and other late essays* (Vern W. McGee trans.; Caryl Emerson and Michael Holquist, eds.). Austin: University of Texas Press.

Carrington, D. (2019, May 17). Why the Guardian is changing the language it uses about the environment. *The Guardian*. https://www.theguardian.com/environment/2019/may/17/why-the-guardian-is-changing-the-language-it-uses-about-the-environment

Dhir, A., Yossatorn, Y., Kaur, P., & Chen, S. (2018). Online social media fatigue and psychological wellbeing—A study of compulsive use, fear of missing out, fatigue, anxiety and depression. *International Journal of Information Management, 40*, 141–152.

Hayhoe, K. (2021). *Saving us: A climate scientist's case for hope and healing in a divided world*. Simon and Schuster.

Keles, B., McCrae, N., & Grealish, A. (2020). A systematic review: The influence of social media on depression, anxiety and psychological distress in adolescents. *International Journal of Adolescence and Youth, 25*(1), 79–93.

Lorenzoni, I., Nicholson-Cole, S., & Whitmarsh, L. (2007). Barriers perceived to engaging with climate change among the UK public and their policy implications. *Global Environmental Change, 17*, 445–459.

Oreskes, N., & Conway, E. M. (2010). Defeating the merchants of doubt. *Nature, 465*(7299), 686–687.

Ramsay, G. (2020). Time and the other in crisis: How anthropology makes its displaced object. *Anthropological Theory, 20*(4), 385–413. https://doi.org/10.1177/1463499619840464

Rushkoff, D. (2014). *Present shock: When everything happens now.* Penguin.

Tsagarousianou, R. (2023). Time and mobility/immobility: The chronopolitics of mobility and the temporalities of suffering and hope in situations of encampment. *Mobilities, 18*(2), 267–281.

Whitmarsh, L., & O'Neill, S. (2011). Introduction opportunities for and barriers to engaging individuals with climate change. In L. Whitmarsh, I. Lorenzoni, & S. O'Neill, S. (2012). *Engaging the public with climate change: Behaviour change and communication* (pp. 1–13). Routledge.

CHAPTER 2

Ten News Values for Climate Communication: From 'Crisisation' to Attribution, Emotional Offsets, Pragmatic Instructions and Punchy Storytelling

The crisisation of events is a common trope in traditional journalism. Media outlets compete for attention and perform their duty to inform the public about the dangers and threats. The primordial nature of news negativity dates back to the times when the society had to be warned about the possible attacks, natural disasters and other issues presenting a risk to the well-being of the community. When the crisisation frame is applied generously and unsparingly, members of the public may react with more consumption of the media—or withdraw.

The theories on uncertainty and media dependency provide a helpful explanation for that phenomenon. People seek information to reduce the state of worry, and they hope to complete the picture of a distressful event with solutions, someone taking over control, creating positive developments (Berlyne, 1960). This information-seeking behaviour is aimed at anxiety mitigation and uncertainty reduction. Uncertainty is deeply uncomfortable for the human beings (Lachlan et al., 2010), and in the digital age they turn to media—journalism, social media platforms, other channels—for tension release or coping purposes (Ball-Rokeach, 1985; Lachlan et al., 2010). The more related the crisis is to the immediate well-being of the media consumers (their habitat, jobs, environment), the more dependent they become on a particular channel or outlet.

© The Author(s), under exclusive license to Springer Nature Switzerland AG 2025
A. Denisova, *Effective Climate Communication*,
https://doi.org/10.1007/978-3-031-67340-5_2

Uncertainty mitigation via media consumption can come at a price to the individuals' mental health. A growing body of research (Ahern et al., 2004; Lachlan et al., 2010; McNaughton-Cassill, 2001; Nolen-Hoeksema, 2000) points to the negative effects that the persistent engagement with the media coverage of crises brings to the readers. These may include posttraumatic stress disorder, feelings of uncertainty, confusion, fear, anger and sadness. The saturation of the media fields with 'crises' of all sorts amplifies the probability that people may develop negative feelings and worry, while continuing to seek solace in the consumption of the negative coverage and updates. Living with constant anxiety is unsustainable—it prevents individuals from engaging with daily tasks and wastes precious cognitive and emotional resources on hypervigilance.

The media companies are part of the issue—'selling anxiety' and 'selling nostalgia' are some of the well-studied tropes of media narratives that have been identified (Rivers, 2008). Pointing to the issues and mobilising people to seek solutions—or buy things and services—to either escape the pressure or mitigate the unpleasant feelings through consumption has been a significant trope of the media industry.

Where does this complicated media context leave us, the generation that has to deal with the all-encompassing problem of climate change? In the environment of saturated crisisation, pertinent frames of pressure and increased anxiety associated with increased news consumption, the media publishers and content creators need to provide a more balanced, empowering communication style to keep the audience engaged with the topic of climate change mitigation and adaptation. This chapter analyses the traditional, persistent news values that characterise a significant proportion of media coverage; it proposes the updated, time-necessitated ten news values that allow for an engaging, audience-respecting coverage of the unfolding events. The chapter delves into the issues with viral communication trends and low attention span, as well as the constraints of the shrinking newsrooms in dealing with the attribution of events to global heating and capacity to report human stories.

One thing this chapter does not do is offering false positivity. It argues for a realistic, pragmatic yet empowerment-focussed media coverage that presents the issues in a clear manner, but also provides avenues for action and suggests stories that document positive developments and human efforts. Climate change is a chronic condition of our planet, not a one-off crisis or emergency, and has to be treated as such.

No Pain No Gain? Why Traditional News Values Are Outdated for Climate Coverage

The analysis of news values across decades is essential to document the shifting understanding of what makes news, and how the notions of conflict and drama tend to dominate the discourse; how the discrepancies between the coverage of domestic events and struggling countries affect the balance of storytelling; and which news characteristics and insights from the news values scholarship apply to the era of climate change.

Much wisdom and hope can be found in the classic and remarkably resonant study of the journalism news values by Galtung and Ruge. The Scandinavian duo of sociologists, Johan Galtung and Mari Holmboe Ruge, came up with the classification of news values in 1965. Since then, their 12 criteria of what turns an event into a newsworthy message have been praised and further adjusted many times.

Ironically, although produced way before the pandemonium of the social media, Galtung and Ruge's (1965) study likened the abundance of media signals to the cacophony, or to the dance of the atoms emitting waves. What they have declared is that there are 12 news values—the conscious and subconscious criteria media professionals apply to choose stories to report to the audiences. Those were:

1. Frequency
2. Threshold
3. Absolute intensity or Intensity increase
4. Unambiguity
5. Meaningfulness
6. Cultural proximity or Relevance
7. Consonance
8. Predictability or Demand
9. Unexpectedness, Unpredictability or Scarcity
10. Continuity
11. Composition
12. Reference to elite nations.

Galtung and Ruge (1965) elaborated that the more criteria an event fits, the more likely it is to be selected for journalistic storytelling. During coverage, the aspects that fit any of these criteria will be emphasised. Some of the most overlooked findings of their study lie in the embarrassing verdict to the foreign coverage—Galtung and Ruge (1965) found that the more distant and poorer countries covered were framed through negative news, or via stereotypes. Galtung and Ruge (1965) also noted that positive news usually refer to the actions of specific individuals while negative news usually refer to natural factors—this observation remains relevant to much of the contemporary climate coverage. The discrepancy between positive and negative news remains complicated in the media coverage— positive achievements are often cherry-picked from the smaller projects of individual human beings, while broader optimistic developments in collective agency do not receive sufficient attention.

It is not by coincidence that another specialism of Galtung and Ruge was the research on peace and empathy. What is worth borrowing from their decades-long work in that field is their 1960s recommendation to the Western newsrooms to cover more 'trivial' news to move away from the representation of the world as full of dramatic events; more space to be given to the predictable and frequent; more experiment with dissonant and distant voices; more interest in the follow-up to the disruptive events to restore the sense of normality. These advices stand tall when it comes to the 2020s climate coverage. Some of the examples of the 'trivial' contemporary climate-related stories that are often overlooked are the scientific innovation that aims to tackle energy crisis or sustainable agriculture; the lives of the farmers and fishermen; the green transition happening steadily in so many parts of the economics and the world.

With the evolvement of the media landscape and the quest for originality, the spike of infotainment affected news values in the early 2000s—the tendency to synthesise serious news and entertainment, sports and gossip, sometimes in a flamboyant order, resulted in the carnivalisation of news, but not the reduction of negative messaging. Harcup and O'Neill (2017, 2001) have been among the most consistent reviewers of the news values across the decades. In 2001, they duly noted the value of surprise factor, celebrity culture, and gave praise to the increasing presence of follow-up coverage, something that must have had pleased Galtung and Ruge (1965). Exclusivity (Schultz, 2007) was another added criterion of value in the age of infotainment.

In the mid-2010s, drama (Harcup & O'Neill, 2017) and conflict entered the top list of the news selection criteria, especially in the British press. Drama in this case can refer even to entertainment stories that may feature celebrities or popular athletes going through some issues in their lives. Conflict has been present in the criteria even before, perhaps more masked by the overall umbrella of 'bad news' (Harcup & O'Neill, 2001) or 'unexpectedness', 'absolute intensity' (Galtung & Ruge, 1965). Yet conflict is a helpful striking characteristic as it underlines the tension that can be seen in politics, culture, society, as well as climate debates— whether it is the argument on sustainable transition, or protests, or political polarisation that accompanies climate coverage in some media markets.

Overall, the news values, in their scientific 60-year-old journey of exploration, have been largely dominated by negativity, conflict, magnitude, proximity or relatability, and much less so by progress, hope and discovery of the best parts in humanity.

Succinctly put, the main news value is... **pain**. The second one is **surprise**—which can come from the position of awe or anger. Ultimately, the current news criteria add an unpleasant layer to the human existence in its current state, raising the levels of anxiety and worry.

Cognitive and Emotional Undercurrents of News Fatigue and Disempowering Storytelling

Too much pain, or even shared pain, achieved through empathy, is propelling many members of the audience into distress. Empathy is a great virtue of humanity, but in high doses it can become debilitating for those with the higher levels of emotional responsiveness and sensitivity to the suffering of the others. Women hurt disproportionately from the emotional burnout of following the news (Villi et al., 2022). (Although, as researchers point out, this might be the outcome of women being more outspoken about their emotions in the interviews than male participants, the finding is still remarkable.) In a vast study of the reasons for news avoidance across countries (from Israel and Argentina to Finland), two main factors have been identified—cognitive and emotional. Cognitive factor means the oversaturation of a specific topic, say, Donald Trump's words and trials, or COVID-19, in the news, while emotional factors have a direct link with the negativity of news. Beckett and Deuze (2016: 2, as cited in Villi et al., 2022: 159) put it tersely: 'the old idea of 'hard' news' that shocks, frightens, disturbs, and alarms can leave the audience feeling alienated, disempowered, helpless and, worst of all, apathetic, insensitive, and even hostile to learning about our world'.

Emotional reactions to the traditional 'hard news' that deal with disasters, violence and problems range from fear and despair to disgust and anger (Villi et al., 2022). The age category of 18–34 is the most affected by the emotional overload and subsequent news withdrawal, and varying cultural and socio-political factors play a modest role in changing that. Across countries, the emotional reasons were strong enough per se to generate news rejection—even if the respondents were coming from the nations with varying levels of media literacy, political stability and media trust (Villi et al., 2022). 'Anybody has a bad start to their day when reading about such things', a 25-year-old respondent confessed about starting to avoid negative news.

In addition to the Generation Z and Millennial categories reportedly being most affected by the news burnout, stay-at-home parents were also at the high risk of news avoidance, sometimes due to the exposure to daytime television, talks shows or pure doomscrolling on social media. They reported feeling powerless and frightened by the negative events reported. This finding echoes the classic study of 'mean world syndrome' by George Gerbner in the 1970s. Gerbner found that the more violent and scary content people consume from the media, the more scared they are about the state of the world, the higher is their distrust in the society. Consuming media violence does not make people violent, it is more likely to cultivate the feeling of fear and anxiety (Gerbner et al., 1986). The distorted and menacing picture of the world—through the media—has an effect on the population that estimates the rates of bad things to be higher than they are and distrust one another.

When it comes to social media, which have dramatically changed the news landscape (Denisova, 2021, 2023), the news values seem to be further reduced to the limited range of negative and positive stories that elicit a strong emotional reaction. 'What is the news?' is the question that is now being answered not just by trained journalists, the newsrooms that may have ideological or commercial agenda, but by the social media owners who use the non-transparent algorithms to manufacture a worldview for their users. These algorithms certainly have a big say in what is being identified as news worth sharing with the millions in a viral manner (Denisova, 2021, 2023). Users do have a part to play, yet they are willingly operating in an environment where news values are no longer set by the media professionals—but by the advertising multi-million companies, which search engines and social media platforms essentially are. The constructed worldview of social media updates relies on clickbait and stories that generate an emotional reaction, at the expense of a more robust and equilibrated argument.

Surprise, relatability, shock and such instinctive emotional reactions as awe, anger and anxiety have been identified as the triggers for sharing (Berger & Milkman, 2012; Denisova, 2021, 2023). Although emotional triggers are not values, they can be seen as such, as they define the habit of the social media users, almost train their reactions, as well as allowing these reactions further train the algorithms—it is a vicious cycle that likely highlights as news value the number of clicks, sharing, discussions around a topic.

Ironically, journalists seek feedback from the audience—or at least seek attention to their stories via online audience measuring tools and social media shares. What people want is a big question, with the classic underlying dilemma whether journalists should give people what they want or what they need. Broccoli or chocolate. When Harcup and O'Neill (2017) examined which news stories from the leading UK newspapers received maximum attention on Facebook, compared to the overview of print news values, they identified entertainment and bad news as the leading criteria. They proposed shareability as the new news value for the social media age, which may entail a striking visual component, a video or multimedia elements, that makes the story worth publishing, even if the information value is low. This book agrees with the shareability concept and expands on it, transforming this criterion into the 'wow effect'. In my previous research on viral journalism (Denisova, 2023), 'WTF, LOL and OMG'[1] were cited by the leading social media editors in the UK as the best predictors of the story going well on social media. In this vicious cycle, these predictors become the very criteria for not just algorithms and social media users, but media professionals when they choose the stories for publishing. Shock, laughter and awe.

Yet another news value that can still give hope to the journalists keen to keep the audience engaged and not lose focus on the important happenings is **'view from the moon'** (as a follow-up on Denisova, 2023)—it is the joker story of quality journalism, more often a feature than plain news. Deriving from a quote from The Economist Deputy Editor, 'view from the moon' is a story that replies to the unasked question 'why' or provides a deeper curiosity to explain events and things that puzzle people. This approach of identifying a strong story takes a step back from the tangle

---

[1] Slang expressions that are decoded as 'What the F***, Laughing Out Loud, and Oh My God.'

of algorithmic settings and goes back to the original values of worthy journalism reporting.

To give an example, the most viral story of The Economist a few years ago was the short video explainer about why Chinese men are struggling to find a wife in a country with the highest population in the world (Denisova, 2023). This short and neat video feature relied on the witty overlay of images and facts and gained an astonishing 3.4 million views on social media platforms, despite not fitting most of the news criteria identified by the scholarship. The 'unexpectedness', 'entertainment' and 'shareability' certainly had a role to play in its triumph, yet it is the soft curiosity and succinct, accessible analysis that rather made the story so successful. A proper 'view from the moon' that no one asked for, but enjoyed when it was provided. A resemblance of a child's curious questions that adults in the room did not think about yet were pleased to know the answers to.

CLIMATE CHANGE AND NEWS VALUES—MORE AGENCY, GOOD NEWS AND MID-LEVEL SOLUTIONS

When the news criteria apply to climate coverage, the exaggerated negativity of the news, documentaries and analyses about the situation does not help the audience's feeling of efficacy. The coverage of climate change falls in the very same traps of the traditional negative news values that have been proven to be unhelpful for the local and distant societies, for the engagement with the socially important agenda, for community cohesion, for the psychological well-being of the population and for the sense of security that is missing and can be abused by populist politicians.

Climate communication needs to acknowledge many difficult truths that lie underneath and make this coverage harder. It is worth distinguishing the main obstacles to engaging the audience with climate coverage:

1. The story is long term and is here to stay for many decades to come.
2. It is multi-layered. Climate change is hard to report in short messages, and the attribution of extreme weather events to the anthropogenic increase in global temperatures is a big challenge of under-resourced and scientifically untrained newsrooms.

3. Most people lack the scientific knowledge to connect the myriads of dots in this story (e.g. explaining how efficient the transition to electric cars in Europe or the US can really be for the big picture).
4. It is mostly negative in the way it affects habitats, seasons, livelihoods, communities, supplies and resources.
5. It is affecting the developing countries more and harsher than developed countries (reporting the suffering of the distant Other has not been known to gain substantial sustained attention in the Western reporting).
6. Solutions are complex and unchecked (technological innovations are in their early stages, natural healing takes a long time).
7. It requires multiple sustained collective efforts (individual human stories are less likely to show real empowerment).
8. Identity threat. In the more polarised discourses on climate change, the adaptation and mitigation activities are equated to sacrifices and identity perils (e.g. when a petrol-car-driving cattle farmer in a remote area in the US is asked to give up their vehicle and use public transport or electric car instead, as well as reduce cultivation of cattle for meat).

In the light of these, negative storytelling and focus on the power elites in solving the crisis, while giving some coverage to the natural disasters and human suffering, seems reasonable in the traditional media logic. Social media logic will arguably call for the apocalyptic imagery, dramatic disaster framing and plenty of entertainment distraction like buying sustainable trainers, to keep people tuning in

Social cultural trauma is another obstacle—the passivity of the governments and their limited ability to bring a consolidated change brings social inertia (Brulle & Noorgard, 2019). Western societies and individuals are asked to change their ways of living, earning and building—and this is another big aspect of a tectonic change that becomes harder to put into practice. The postcolonial perspective suggests pointing to the privilege of even having this choice—of acting or not—while many countries of the Global South are already taking a big hit to their ways of living due to climate change. Yet this book is pragmatist in its viewpoint that it is an unreasonably lofty expectation to change the way the wealthier countries see the distant 'other', and how historical injustice has to be rectified. This book is about acting now, making sure more damage is not done to the suffering communities across the globe, and conversations about the guilt and shame have to be treaded lightly not to fuel new populist movements that risk overthrowing the subtle yet growing global cooperation on climate.

Yet the alternative perspective on cultural trauma emphasises the role of the narrative to either construct or minimise this trauma (Alexander, 2004). It is down to the populations, often directed by the governments, media and influential social leaders—to explain the events and frame them either as a tectonic change or gradual organic adaptation; as a new cultural and societal system; or as the evolution of the existing one. Not every failure or crisis generates trauma—only in those cases, when a society equates it to the cultural crisis, a threat to identity.

While avoiding trauma might be the natural societal urge, Alexander (2004) reminds us that rejecting the notion of trauma and working through it means ignoring the suffering of the others, denying other societies their traumas. From this perspective, the collective feeling or way of thinking about the climate change may or may not fall in the definition of cultural trauma. In the media, construction of climate change as a trauma much depends on the geographical location, resources and means of mitigation and adaptation.

The conflict—or at least tension—exists between climate reformists (those in favour of steady reforms) and climate radicals (those arguing for the immediate climate justice and the overhaul of societies and institutions) (Hadden, 2015, as cited in Brulle & Noorgard, 2019: 6). The reformist approach relies on Green Governmentality and Ecological Modernisation (Brulle & Noorgard, 2019). These comprise strong international governance, including investment in green technology, better use of resources, and tax incentives and tax punishment.

Climate justice, on the other hand, derives from the tradition of challenging the capitalist hegemony and asks for the structural changes that in turn should lead to the green overhaul. This radical approach points to the Global North/ Global South divide, systemic inequality and the critique of the current international governance (Della Porta & Parks, 2014, as cited in Brulle & Noorgard, 2019). Climate justice pioneers accuse climate reformists of adopting the post-political climate stance, assuming that capitalism and market economy is the default setting of the humanity, and all climate actions have to arise within this framework (Swyngedouw 2011: 264, as cited in Brulle & Noorgard, 2019: 6).

How can journalism and climate conversations on social media overcome this ideological divide? Some commentators see the debate between the two fractions as the source of delays to climate action. Others argue that, without deciding on the underlying strategy, tactics make little sense to mitigate climate change. Political sociologists caution that both sides accuse each other of social inertia, inhibited action over assumed responsibility of action. The reformists wait for the governments and businesses to act, while radical climate activists call for the reformation of the social system before climate strategy can be put into action.

Journalism is stuck between the rock and the hard place—do professionals need to pick a stance? How do they choose the events to report on, and which developments should be highlighted for the significance in climate mitigation and adaptation?

In this light, to liberate journalism from the ideological dilemmas, the news values adapted for climate coverage are proposed. The laconic list is presented first, with the more deliberation for each value provided afterwards:

1. **Bad news**—major natural disasters will surely occupy the headlines, as the effects of climate change exist and need to be shown to the publics worldwide.
2. **Frequency, but told from a new perspective**—the increased probability of natural disasters is highlighted in the news more and more often. I am arguing that the accentuation of the changing patterns deserves its headlines. Instead of focussing on single heartbreaking events, the *pattern* has to be emphasised. This angle will put things into perspective—and provide framework of living for the audience. For example, for now unexpected heat or unexpected cold spells are treated as one-off cover pages—they deserve to be covered systematically.
3. **Positive action/good news**—the initiatives of the governments and intergovernmental organisations may sound formal, broad, boring and written in bureaucratese, yet they are the stem to the emerging big developments to tackle the crisis.
4. **Emotional offsets**—there should be more effort to identify 'light and shade' in climate communication to break the stereotype of climate stories being the 'shade' and other themes providing material for the 'light'.
5. **Innovation**—technological and natural innovation needs more spotlight. The caveat here lies in becoming overly optimistic about tech-solutions and generating the risk to create more inertia, yet

more coverage of the projects happening is likely to increase the sense of hope, security and agency in the audience. Over 1 billion dollars are being poured into green transition by the US government, US investors, Chinese government, European Union and private investors—some of these projects will be a fiasco, others will become the new 'Apple', 'Tesla' and 'computer' of tomorrow. Solution journalism can be a variation of that. This beat should be accompanied with scrupulous investigative journalism—to hold powers to account and monitor the instances of fake solutions and greenwashing.

6. **Awe**—the surprise effect that stories about natural beauty or nature restoration generate in the users are here to stay. One can argue they further help to reconnect humans with nature and remind them of the grandeur of our habitat and its healing properties.
7. **Cultural relatability**—this factor can help 'bring home' stories from a few kilometres away, or from the very distant communities and countries. Finding the common grounds, the way to bridge distant suffering or distant innovation and cleverness with the local audience is a fantastic challenge for climate journalists.
8. **Entertainment and lifestyle novelties**—although it may seem a light topic, soft news and entertaining coverage is key to keep the climate storytelling part of our everyday lives, not a special effort on a good day. Lifestyle coverage and relatable celebrities who care about climate have the power to keep climate action on the agenda of average readers. Smaller individual actions may seem meaningless in the grand scheme of things, but they are not—social psychology proves that they bring us closer to pro-climate identity. There is a small limitation though which I call 'greenwishing'—more on that in Chapter 4.
9. **Mid-level activity and solutions**—as opposed to the 'lone hero' or 'technology will save us all' narratives, this approach proposes having a close look at the collectives that make the changes that matter—communities, businesses and industries that come together to tackle climate change one issue at a time—and achieving results at this. There is a tendency in the Western media to report on the power elites and on dramatic human stories, but the massive volume of pro-climate activity and progress in the middle is often left to the nice business or charity publications.

10. **Empowerment**—in all forms, be it voting decisions, consumer decisions or non-decisions (about no flying, for example), and institutional and business changes that react to the climate threat. We are social creatures, and we need social reminders of the positive activity of the others to compare with our own. We need stories of agency from all parts of the world—the troubles of the distant communities need to be reframed in a new light, accentuating humanistic proximity, common lifestyles, stories of nerve and perseverance, stories that trigger empathy and solidarity. Stories that show communities re-organising and curbing the climate problem.

Let's explore the validity of these climate-conscious news values one after another.

1. **Bad news.** The respected scholarship on news values, from the mid-twentieth century and until the present day, proves that negative happenings will always be in the news. The duty to report troubles and issues is the primary responsibility of journalists—it rings the alarm bells for the population that may need to defend themselves, adapt or take other measures. An American gatekeeping theorist Professor Pamela J. Shoemaker (1996) calls it a direct evolution of the survival instinct—people are genetically programmed to monitor the environment for the signals of potential dangers. She follows the metaphor of Lasswell (1960, as cited in Shoemaker, 1996: 32) who compared journalists to the members of some animal species who keep the guard for the herd and alert it to any danger.

   While the potency of the survival instinct is not to be overlooked, the veneer of civilisation and culture arguably allows humans to assess the situation beyond the local and distant alarm bells. The research shows that humanity is living better than ever—there are fewer diseases, crimes, people are living longer and more fulfilling lives (Rosling et al., 2018). The 'fight or flight' response is not a sustainable reaction to news triggers, especially for those lucky enough to live in the places that possess relatively modest threats to the existence—not all places on Earth are managed this way, of course, but those who can benefit from the calm progress and deliberation of ideas can use their cognitive and mental capacity to keep the progress developing, rather than losing sleep over bad news. The Swedish physician, Professor Hans Rosling argues that we are overwhelmed by the feeling of dread, while the facts speak the opposite—we have more reasons to be cheerful and optimistic than ever before.

'Pathologies of information' is a curious term to define the overwhelming quantity of information people are exposed to—from both the traditional and social media. Even before investigating the many influences of social media on well-being, understanding the traumatising effect of watching violence on television brings the fundamental comprehension that secondary trauma exists—the increased viewing pattern of traumatic televised context increases the feeling of uncontrolled fear, psychological hyperarousal and sleeping difficulties (Bodas et al., 2015). Viewers consuming news bulletins regularly are 1.6 more likely to report at least one of the anxiety symptoms mentioned before. Not everyone would react the same to the upsetting content—yet an increased media consumption of negative news intensifies anxiety even among the more rational viewers (McNaughton-Cassill, 2001), and those with the developing burnout will suffer the most (Havrylets et al., 2018).

In a review of the many studies on information overload and information anxiety, Bawden and Robinson (2008) have discovered several pertinent issues that affect the quality of news engagement and well-being of the audience. Firstly, they defined 'information overload', an elusive concept, firmly, and linked it with the abundance of media messages:

> The term is usually taken to represent a state of affairs where an individual's efficiency in using information in their work is hampered by the amount of relevant, and potentially useful, information available to them. (Bawden & Robinson, 2008: 182–183)

To put things into perspective: a weekly edition of the New York Times contains more information than the average person would read in a lifetime in seventeenth-century England (Bawden, 2001). Given that this estimate dates to the scholarly piece produced two decades ago, one may only wonder where the comparison stands for the 2020s.

Secondly, the issue of the loss of authority, micro-chunking and shallow novelty have been identified as the caveats of information on social media. Users pour their precious mental energy and attention span into engaging with fragmented, often unimportant, re-packaged, de-contextualised or exaggerated rudiments of information and analysis. This leads to them struggling to identify the authority or reliability of those, which concludes in the build-up of stress. (Bawden & Robinson, 2008)

Another 'pathology of information'—information anxiety—adds to the understanding of what happens to human cognitive systems when overloaded with data. Wurman (1989) describes 'information anxiety' as a stressful condition of not being able to access, process or act upon the information consumed. It results in 'environmental numbness' (Gifford, 2011) when an individual becomes numb to the overload of climate information, especially if this information is too frequent and not too varied in tone and content. Information anxiety comes with the side effects of reduced focus and diminished confidence (Renjith, 2017). The metaphor of 'infobesity' is also helpful as a quirkier term to point to the toxicity of the excessive consumption of the information of dubious quality—it brings brain fog and the unpleasant 'digestion' afterwards (Morris, 2003, as cited in Bawden & Robinson, 2008: 185).

When information is mostly negative, people are more likely to withdraw or process it less efficiently. Some of the cures for information overload include taking control of one's information channels, adjusting times of the day to engage with it, applying critical thinking and mindfulness to detect the effects that the pathology of information has on one's body and mind. Reducing the negativity of the stories consumed is another method of lessening some of the stressful consequences for media consumers.

Surprisingly, some degree of worry and anxiety can lead to the increased sense of efficacy. A study of Italian university students demonstrated that the two can co-exist in response to the climate crisis (Maran & Begotti, 2021). However, the intensity of anxiety is an important factor here—it has to be moderate enough not to overwhelm the cognitive system and leave room for creative and rational thinking.

Media producers are starting to acknowledge that the unquestioned focus on the negativity is not a default. Craft and Wanta

(2004) found an interesting variation between the proportion of negative, neutral and positive news between the female- and male-led newsroom. Those with women at the helm tend to report negative stories less often. Female editors were also found to prioritise social and humanitarian issues over political news or catastrophes, thus reshuffling the dynamics of the traditional, male-dominated media teams. Female editors in chief were found to encourage their teams to focus on the positive aspects of stories (Craft & Wanta, 2004).

Can the media stop publishing bad news about climate change? Of course not. Yet being more selective in choosing what to publish, at what frequency, in what context can assist in reducing the information anxiety effect. This will likely keep the audience engaged and, in the long run, empowered.

2. **Frequency, but told from a new perspective.** 'Global weirding' (Hayhoe, 2021) is a more correct moniker to underline the effects of climate change on the planet than 'global warming', a slightly outdated term from the last century. Temperatures rise and fall, and extreme natural events occur at a higher frequency and following the yet undetermined pattern. This is a major risk—humans can no longer predict when their homes will be cold or warm, flooded or affected by forest fires.

    Extreme weather events are incredibly costly to human lives and to the infrastructure—preparation and mitigation is way more humane and economically viable than dealing with the consequences of disasters. Media must serve as an advocate for the population that is likely to be affected by the extreme weather incidents—journalists should receive appropriate training to be able to explain the emerging patterns of 'weirding', to link global disasters to the issue of the disrupted planet. In this framework, the frequency and intensity of events—a classic news value—will be better integrated in the climate action paradigm. It will help the audience to connect the events and—in the fragmented and overwhelming info-noisy environment—to learn the clear narrative of the climate change domino effect.

    Up until the last decade, the unhelpful media logic of perceiving climate change as a matter of debate, rather than a confirmed happening, has twisted the coverage towards discussion and uncertainty. The BBC stopped platforming those challenging the anthropogenic nature of climate change only in 2018. The US media have been devoting plenty of space to the voices from, what seemed at the time, all sides of the matter. The allegiance to fake balance—and interest in conflict, polarisation—has stalled the progress of climate communication and action for decades (Boykoff & Boykoff, 2007). This has also slowed down the literacy and understanding of the cause-effect relationship between climate transformation and extreme weather events.

The good news is that the media coverage in the West has moved beyond the false balance of giving room to both climate scientists and climate deniers. However, the emerging understanding of the expanse of climate change has a name—it is called **climate attribution** (Hopke, 2020), the indirect and elusive link between extreme weather events and climate change. Can all wildfires be attributed to climate change? Is hot August in Europe always indicative of the poor state of the warming planet? With the challenges in determining the effect of climate change on specific developments abound, journalists usually err on the side of caution—unless their publication is of activist, highly ideological or well-resourced nature. The lack of specialist knowledge is understandably common and prohibits media professionals from making big conclusions (Engesser & Brüggemann, 2016; Wilson, 2000).

Aligning forces with scientists to include a systematic evaluation of the climate change influence might be a step forward in overcoming this hurdle. In fact, such Western media as Carbon Brief prides itself in having an allocated team of scientists who serve as advisers to journalists and are available for fact-check and commentary at a short notice. Media outlets have been slow to follow in this trend, yet climate attribution is slowly gaining momentum in the heatwave stories (Hopke, 2020).

The challenge number two is the unequal distribution of the awful and not-so-awful events across the planet. While it has been proven multiple times that climate change is destructive for all in the long run, some local ecosystems may actually benefit from the effects of global warming in the short term. A highly cited paper by 30 climate scientists (Meehl et al., 2000) gives credit to the improved bird proliferation in some parts of Australia, where birds prefer to breed after heavy rainfalls; another example is the hundreds of lives preserved due to milder winters in the US, where several states that usually suffer from extremely low temperatures have had lighter winter seasons. Optimistic as these discoveries sound, they are not recommended for headlines—these cherry-picked reassurances are better replaced by the clear pattern exploration of the climate-affected frequency and intensity of the weather events.

The very same scientific paper of the American Meteorological Society (Meehl et al., 2000) makes a less reassuring point that while there are measurable characteristics in the climate change science (spikes in temperature, the duration of extremely hot or extremely cold days, disasters intensity and occurrence), the relationship between human society, flora and fauna and the changing climate 'is not intuitive'. They point to the changes in population density and vulnerability in specific areas on Earth. If the hurricanes of the 1940-50s occurred in the present day, their effects would be worse than in the last decade—all due to the changes in human exploration of the areas, activity performed in those and more people living near the coasts. More adaptation to the extremes can help reduce the climate impact in some areas—and science-backed journalism can help advise populations in context.

The good news is that scientists and governments are rolling their sleeves up to develop more accurate models and monitoring technology to make the links between the events more acute, to empower the forecast technology and to educate various fields of science—from agriculture to medicine—about the relationship with climate change. Journalism should be among the first recipients of the developing climate attribution training.

The upward trend is evident—and so is the resource gap. More and more media outlets mention climate change in their coverage of extreme weather events (Hopke, 2020). Yet the confidence in climate attribution is more common for wealthier national titles or better-resourced climate publications—The New York Times, Washington Post, NPR, the BBC and the Guardian along with Climate wire were associating heatwaves and wildfires with climate change with greater frequency than regional or non-climate-focussed outlets. It has been established that the publications that invest in climate specialists and climate teams in the newsroom tend to report the extreme events for longer and to evaluate the climate attribution with more frequency (Hopke, 2020).

Front page splashes about climate change accelerate people's worry and anxiety—but do not create a sustained interest. Placing a climate story consistently within a media publication—not necessarily on the front page but dedicating effort and interest to it, demonstrating the pattern that it makes a part of—can help sustain climate change awareness.

While elite publications set the tone, they are clearly early adopters here. The growing presence of climate reporters and editors in media publications is a sign that more investment—or redistribution of resources—is needed in other newsrooms. They can either train the existing staff more to make sure climate data and climate science are understood fuller and in relation to every beat. It is a necessary step in innovation and adaptation—to enable the journalists to make links to climate science and acknowledge the climate attribution of the higher frequency of extreme events.

3. **Positive action/good news**. Reporting action being taken to tackle climate change is paramount to the long-term management of the crisis. We need to see examples of various levels of action—from the supranational and intergovernmental action to the governments, industry and citizen collectives mobilising for the social good. As important as climate agreements, reports and conferences are (the likes of IPCC, COP and similar), their coverage faces a challenge of translating legal and scientific language into terms relatable to the general audience. Such climate media project as Carbon Brief in the UK is a solid example of user-friendly yet high-level climate journalism—funded by the EU, on this project journalists and scientists work alongside to produce clear and accessible climate stories, updates and explainers.

Can reporting good news have a negative effect on mobilising citizens towards pro-climate thinking and acting? Yes, if the beliefs are not steady and can be tweaked based on the incoming information. Thus, Sunstein et al. (2017) warn that both good and bad news on climate change can generate varying responses in the audience, depending on their beliefs (see also Shanahan, 2007). Those with the strong belief that climate crisis is a man-made problem respond more to bad news—it reaffirms their values and makes them more committed to keep finding ways to address the issue. On the opposite side, those still questioning the threat of the climate crisis are more affected by good news—these serve as the reassurance that the crisis is 'overrrated' and would dissolve on its own. This recent study (Sunstein et al., 2017) demonstrates how climate communication can be polarising for various groups of people—and that it can possess inherent tools for populist politicians to divide people along the lines of climate beliefs (Hayhoe, 2021; Sunstein et al., 2017).

Nonetheless, the quantity and content of those good news stories matter. While Sunstein et al. (2017) often use an example of the adjusted climate estimates as an example of good news (e.g. 'a particular region is heating up slower than most scientists thought'), there is a whole pool of positive stories to report. These range from the historical agreements and deadlines for their implementation; legally binding commitments (such as the UK's 2050 net-zero target—it is a law); ground-breaking limits on fossil fuel production and expansion; tax credits and green investment initiatives (Rathi, 2023); industries adapting and pivoting to green; small business stories; agriculture inventions; community adaptation and thriving; and so on and forth.

Naturally, the survival instinct does not get blood pumping to the discovery that …elephants prefer eating leaves from the trees that are not storing carbon, thus contributing to the sustained growth of the carbon-capturing trees (Millan, 2023). The announcement of the new public transport infrastructure may also be met with mild enthusiasm—nothing compared to 'which species will be extinct by which year' colourful interactive infographic. The latter is likely to cause awe and anxiety, much stronger reactions than mild amusement and a light feeling of hope triggered by the former stories.

Negative information, repeated constantly and consistently, has more influence over impression formation than positive one. People respond to negative stories more strongly, and these are shaping their worldview faster—what Soroka (2006) calls the asymmetry of information and effect. When the media persistently report negative information about a topic, the public perceives the story even more negatively overall. The continuous negative coverage of a difficult story (e.g. economic crisis) can deepen public pessimism by as much as 16% (Ibid).

What I am arguing for is the increased proportion of good news and representation of positive action in the overall media coverage. Bad news is not going away—but it should give space to other, more progressive and reassuring stories that reflect more truly various developments on the ground.

Only 25% of stories about climate in the UK newspapers are positive (Futerra, 2006). The frame of 'catastrophe' in relation to the IPCC report still dominates lots of media coverage in Brazil, China, India, Russia and South Africa (Painter, 2007); negative storytelling is the prevailing *modus operandi* when it comes to the media decisions on climate reporting (Shanahan, 2007).

Yet the audience needs more positive news to offset the information asymmetry. It also shows the signs of actively *seeking* more positive news, perhaps demonstrating the other side of the phenomenon that is survival instinct. Unable to cope with the spiral of doom, people tune in to healthier and happier stories. Around 60% of Twitter users who read journalism online prefer good stories—a discovery that has been measured through clicks, replies and shares; and only 47% actively engage with bad news (Al-Rawi, 2019). The gap becomes wider on YouTube—over 77% favour good news, and 23% engage with bad news on the video-sharing platform (Ibid.).

4. I am proposing the concept of **'emotional offsets'** in media coverage—for every negative story there should be one story that looks into solutions, collective action, positive initiatives, inventions, intergovernmental agreements, among other ideas. The emotional responses to news can impede the sense of control in the viewers (de Hoog & Verboon, 2020). While not everyone reacts in the same way—some will feel the burden of the world on their shoulders, other just brush the news headlines off—there is a tendency to feel negative emotions after the exposure to news headlines. The main route to mitigate the negative perception of the news is to give more sense of control or present less severe happenings in the stories (de Hoog & Verboon, 2020). It is obvious that diminishing the effects of climate change on the world is unethical and untruthful; however, offering more stories that can boost efficacy and the sense of control is a responsibility of the media organisations that are keen to keep the readers engaged, not withdrawn. This is why the variety of climate news is encouraged, comprising

those that upset and those that can show the avenues of improvement. Some of the improvements that newsroom could implement is integrating more constructive journalism into their agenda—the stories that provide the information about how people can cope with the distressing events; it provides a variety of perspectives and ideas; it is empathetic yet impartial and forward-looking (Bonn Institute, no date).

It may not sound easy at first, but good journalists never shy away from a worthy challenge. The rise of solutions journalism is a considerable achievement of the 2020s, but it can be accompanied by other types of positive climate storytelling, and a strategic action in this direction can help to bring more balance to the currently overwhelming and negative climate coverage.

5. **Innovation.** Stories of innovation are exciting—isn't it marvellous to see the rise of the new Einsteins and Newtons, new vaccination programmes and rockets that land back after flying into the outer space? Innovation is crucial for growth (Simpson & Tamayo, 2020). In case of climate change and its triggers—fossil fuels, human activity—a reduction in the use of resources and an increase in productivity is a much-desired goal.

An enthusiastic and persistent coverage of new technological developments may be accused of technocracy or capitalist thinking, or it may lull the audience in complacency. A heightened media focus on technological progress and scientific discovery diminishes the calls for individual and community mobilisation, offering to the citizens the false assurance that 'technology will save us all' (Molek-Kozakowska, 2018). However, we are beholding so many new technologies and inventions emerging in real time—we may as well get excited by bearing witness not just to the colossal damage done to the planet but by the gift of intelligence and natural resources to apply to clean energy. 'The world is facing two industrial revolutions, one powered by artificial intelligence, the other by clean energy', Shrimsley (2023, para 3) argues. Looking at the innovation and the advantages that come with it results in understanding of 'the economic benefits of being a pioneer in an industrial revolution which cannot be avoided' (Shrimsley, 2023: para 15; see also Rathi., 2023).

Who would not want to discover the big new breakthrough? Yet, for businesses and institutions at the forefront of innovation, media exposure is favourable and damaging at the same time—and here lies the challenge of media coverage as a double-edge sword for innovators.

A media spotlight on innovative developments and start-ups can help inform the market on the activity of a company; reassure investors of its validity; and bring more money to the table (Milbourn, 2003, as cited in Dai et al., 2021: 345; Simpson & Tamayo, 2020). Yet it can also put too much short-term pressure on achieving quick results and alert the competitors to the specific details of the technological advancements of a company (Dai et al., 2021). An additional perspective needs to be considered too, especially by the critics of capitalist society—the media surveillance helps to curb the greed of the managers and ensure they invest more money in innovation and expansion rather than lavish bonuses; greenwashing claims are likely to be interrogated in the public eye (Dai et al., 2021).

The first challenge in covering innovation lies in the incredibly inflated vocabulary around 'innovation', 'novelty' and breakthrough', exhausted by both the attention-grabbing newsroom sub-editors and the marketing and public relations industries at the service of the commercial innovator companies. Very often the label 'innovation' is put forward by the market players when it benefits their needs, not when the real discovery is at the heart of the story (Mas, Huck & Zerfass, 2005). What gives reasons for optimism is that the word has not lost its exciting connotations—journalists (see, for instance, Mast, Huck & Zerfass, 2005, on the German media scene) still evaluate the term positively; they see little risk in innovation and tend to assess innovation claims critically.

The second challenge for innovation reporting is that newsrooms may not have sufficient expertise to weigh every new project coming from various industries. There are media companies that reach out to the experts in the field to help them evaluate the merits of the new discoveries, yet this approach is not bulletproof and often requires more resources than a simple fact-check and subsequent publication of an edited press release.

The third challenge is that the companies themselves rarely provide the full description of the project, aware of the competitors hawking around. This limited transparency adds another layer of difficulty to the media professionals trying to assess the meaning, efficiency and long-term perspectives of the innovation (Mast, Huck & Zerfass, 2005). Some journalists argue that technological innovation is better left to the specialist publications—this, nonetheless, inhibits general population from learning about the promising projects in the making.

The fourth—and easily the most head-scratching—challenge of innovation coverage lies in dissecting the overpromising corporate reports and social responsibility claims (more on this in the chapter on greenwashing). It is a massive 'green' elephant in the room—companies go to the extremes to come across as sustainable as this creates tax breaks, incentivises investment and improves credit score. There are real benefits in claiming one's business as innovative and sustainable—and a very limited set of instruments in the hands of the media to scrutinise every declaration. Bloomberg's 2022–2023 investigations[2] into fake carbon offset proimises set a high bar in this increasingly influential beat of journalism. It requires time, knowledge and investment in the newsrooms, as well as the support of a robust legal team, to pursue an investigation of the sort.

Focussing on the real-life impacts of the innovations—and reporting human stories in a broader context—is a helpful strategy to reduce the fossil fuel inertia. A powerful example comes from Poland, an Eastern European country that relies heavily on coal. Patryk Strzałkowski (2023), head of the Green Desk at Gazeta.pl, recommends showing more solutions at work, as this helps to beat polarisation and climate news fatigue. He points to the example of the reporting that demonstrates that almost as many people are now employed by the electric vehicles industry in Poland as by coal (56,000 against 80,000). This documentation of the shifting social dynamics helps to improve the audience's understanding of the social benefits of green transition, as well as inform them of the green innovation taking place.

---

[2] Rathi, A. (2022) Inside the Billion-Dollar Market for Climate Offsets, Bloomberg Green. 21 November. Available at: https://www.bloomberg.com/news/articles/2022-11-21/junk-carbon-offsets-allow-companies-to-claim-they-re-carbon-neutral?leadSource=uverify%20wall.

Elgin, B., Marsh, A. and de Haldevang, M. (2023). Faulty Credits Tarnish Billion-Dollar Carbon Offset Seller, Bloomberg Green. 24 March. Available at: https://www.bloomberg.com/news/features/2023-03-24/carbon-offset-seller-s-forest-protection-projects-questioned?leadSource=uverify%20wall.

6. **Awe**. The goddess of beauty, Aphrodite, arose from the dazzling pearly sea foam in the turquoise waters near the island of Capri known for its white cliffs and tranquil views. Stunning as she were, she inspired nothing but the feeling of awe in those who saw her—at least that is what the Greek myth tells us. Some converted this awe into servitude, others into lust, yet the major reaction was a variety of forms of admiration, interest and fascination. If Aphrodite were to appear from the warming Mediterranean waters these days and the video made rounds on TikTok, it is very likely that it would harvest thousands of clicks, comments and reposts—and go viral. Not by chance Al-Rawi (2019) and Berger and Milkman (2012) noticed that 'awe' is one of the strongest predictors of a journalism story going viral on social media.

The feeling of surprise and wonder—awe—has been used to remind of the presence of God (awe at the natural beauty aids our spiritual journey, argue de and Luca-Noronha, 2021). It is the emotion situated *'in the upper reaches of pleasure and on the boundary of fear. Awe is felt about diverse events and objects, from waterfalls to childbirth to scenes of devastation. Awe is central to the experience of religion, politics, nature, and art'* (Keltner & Haidt, 2003: 297).

Awe has been studied through the prism of religion, sociology, arts, among other disciplines. Max Weber famously discovered that in the times of political turbulence voters stick to the charismatic leader even if this decision is largely irrational. Charisma, therefore, can be a source of awe (Weber, 1989, as cited in Keltner & Haidt, 2003: 299). Emile Durkheim corrected this by assuming that individual and group emotions can be powerful for mobilisations, political activation and transformative action. Not just charismatic leaders, but the strong structure of feeling can sustain an uprising, or collective action, for the benefit of many (Durkheim, 1887, as cited in Keltner & Haidt, 2003: 299–300).

Piff et al. (2015) bring even more reassurance to the collective fuel that is awe—they state that the experience of awe reduces individualism and aids prosocial behaviour. The 'feeling of small self' is the result of experiencing awe—it puts things and humans into perspective and motivates selfless action. Keltner and Haidt (2003) agree with that by underlining that awe is inseparable from vastness—a picture of a natural wonder may be larger than life, and so can be the acts of a charity leader, or the scale of an immense mathematical equation—and human cognitive reaction to this vastness is strong. A person may need to slow down, feeling spellbound and struggling to accommodate it. By accommodation, theorists mean adapting the existing mental structures and view of the world to the new discovery, the aptitude to make sense of this immensity and power.

Psychology scholars are more pragmatic—they remind that awe can be pleasant or devastating (Lazarus, 1991). It is an ambiguous emotion that can dip into light as well as into darkness. Shiota et al. (2007) postulate that not every encounter with awe calls for accommodation—sometimes people stay unbothered by the experience. However, most studies agree that awe requires a certain—even subtle—response from the human systems, and it is most likely to be the acknowledgement of the small self with respect to the grand phenomenon, with the prosocial behaviour and connection with the others and with divine powers abound.

The stories about climate change that elicit awe are a strong trigger for engagement, self-appraisal, collective re-evaluation and the aspiration to do good. The research from cognitive scientists, psychology, sociology and philosophy demonstrates that the enchanting hybrid of power and wonder (that awe is) is irresistible to our clicks, but also transformative for our ways of seeing the world and our role in it.

7. **Cultural relatability.** *'Why are social media buzzing with the pictures of New York covered in fog? Doesn't anybody care about the Canadian population put at risk by the heavy wildfires happening in Canada and sending ash to the neighbour?'* That is the conversation I had on a very hot day in June 2023 in London, during a one-day workshop between academics and public service media professionals in London.

Many colleagues were wondering about the impact of imperialist thinking, colonial legacy, cultural insensitivity, media hegemony and news values when it came to the reports about Canadian wildfires that sent ochre fog to the streets of Manhattan. Many global media prioritised the images of New York in their front pages. There was something ethically wrong, a clear unequal weighing of the impact of the same events on two spaces at the same time—the Canadians were receiving more damage than the US citizens yet the iconography of New York under a tawny veil of fog was hard to resist for media publishers.

New York in smoke, as previously Notre-Dame-de-Paris cathedral in flames a few years ago, is a massive intracultural symbol. Cultural domination is true and has its evils, yet what newsrooms were dealing with in both cases was a huge recognisable cultural symbol losing its properties.

New York has been enthroned as the peak of the contemporary cultural world by the myriads of Hollywood films, books, cultural essays, magazines like New Yorker and Vanity Fair, cartoon princesses and Home Alone astute boy in Central Park, Anna Wintour and Vogue, Met Gala and rap music. In most parts of both hemispheres, people would be able to say what New York is and what cultural reference they associate with it. Would the same ring true for Canadian rap music, glamourous events and glossy magazines? Arguably the pool of people who have immediate references would be smaller. The same stands true for Notre-Dame-de-Paris, a Medieval cathedral in Paris that had been immortalised by the novel or Victor Hugo, and the subsequent films, plays, cartoons, merchandise (!) and the long-running Les Miserables (or Les Mis, as the musical is known in London West End). Not many of other humankind's cathedrals enjoy the same level of publicity.

Is it right or wrong that only a privileged few landmarks or locations get into the spotlight of the global media? Wrong, from the inclusivity and diversity perspective. Wrong, as the proof of cultural hegemony and West-centric media interests. Yet can we be pragmatic about it? The current challenge is to attract the audience to climate change information and use all tools available to do so efficiently.

In the classic news values taxonomy, cultural proximity is one of the key criteria. Cultural proximity is the sense of relatability to the culture, lifestyle, ways of thinking and being depicted in media representation. It builds on cultural capital, not just identity (Straubhaar, 1991, 2014). In earlier literature, cultural proximity was defined by the distance between countries, the exchange of goods and services, tourism and similarity in political systems (Straubhaar, 2014, see Trepte, 2008 for reflection), as well as common language

(Straubhaar, 1991). Yet, cultural proximity in the 2020s clearly overflows beyond these rigid physical categories and can manifest in the sense of attachment, longing and even projections.

For instance, Zaharopoulos (1990, as cited in Trepre, 2008: 4) noticed that, during the US Presidential Elections, Greek newspapers were covering the American-Greek Democratic candidate much more keenly than the other leading candidate, George Bush. Another example is the curiosity towards foreign-made TV series and shows on the streaming platforms like Netflix, Amazon Prime, Disney+—contemporary audiences are willing to ignore the hurdles of reading the subtitles or watching a dubbed cultural product; they feel they can relate to the extraterritorial characters and encounters occurring on the screen (Limov, 2020).

Back in the 1980s, hegemonic countries would dismiss foreign cultural production under the term 'cultural discount' (Hoskins & Mirus, 1988, as cited in Trepte, 2008: 5)—this meant loss of audience and interest in the markets different from where a show was produced, due to the unwillingness to appreciate dubbed dialogues and unfamiliar situations. In the 2020s, though, 'transcultural affection' is coming into play, signifying selective localisation of the elements of the foreign cultures seen on the screen (Ju, 2020). One could argue it is a cultural proximity spiced up with cultural exoticism and self-discovery tools for the audience. Berg (2017) adds to this by highlighting those complex identities of the modern generations—being raised with access to various media channels, sometimes relocating for work and education, hybrid associations and sense of belonging to various spaces at once—open more avenues for cultural association and exploration of the others as well as of the self.

What it means for the news values is that cultural relatability can manifest in race, gender, class, family status, lifestyle, religion and similar practices—due to the globalising effects of both transnational and social media, there may be more avenues of cultural relatability. To give an example, a frontline health worker, a grocery shop assistant and a pharmacy worker became highly relatable figures during Covid-19 pandemic (Phillips et al., 2023). They were often represented as the victims and heroes of the stampede of worried citizens or panic buyers. Many videos of Chinese health workers circulated in the Western social media, thus bridging the gap between cultures of varying ideologies and political order.

Student population is another example of a universally relatable bunch. Commonly characterised by curiosity, persistence, messiness and humour, the student cohort has sparked many memes that speak across cultures (Ask & Abidin, 2018).

Using professions, collectives and frames that are universally relatable can help reduce the 'othering' of the climate-related human suffering in distant countries. It can show the universality of human experience, and the impact that climate change is having on funny students, hardworking nurses, geeky solar panel engineers and other professions varied by class and status in faraway lands. Personalisation, or turning coverage into human stories—showing how an individual experience is emblematic of a broader social issue—is another classic tenet of journalism (Wahl-Jorgensen, 2013).

Linked to this is another technique of relatability—emotional involvement. This can be achieved through emotional content or emotive language in some reporting, especially closer to the feature writing, less in news (Peters, 2011; Tenenboim-Weinblatt & Baden, 2018). This technique is also known as what Professor Karin Wahl-Jorgensen (2013) calls 'the strategic ritual of emotionality', and it relates to long-form and literary journalism in the more quality publications, as well as sensationalist daily coverage in tabloids. This technique has a pitfall of turning human drama into thrill-seeking of the audience—accustoming the readers to the emotional voyeurism, or 'climate porn', as Ereaut and Signit (2006: 4) unambiguously define it.

Overall, the balance between carefully curated cultural relatability and exploitation of drama and emotive storytelling is subtle. Yet, as early scholarship on attention to environmental problems (Downs, 1972: 38) reminds us, an 'issue attention cycle' in the media is short, and a creative array of techniques to keep it afloat is essential to sustain the audience's interest.

8. **Entertainment and lifestyle novelties.** Once in a small independent bookshop in Brick Lane in East London, I saw a section that combined 'culinary/climate' books all in the same bookstand. What sounded frivolous at first, struck me as a playful take on climate conversations—while not all climate solutions end in veganism, it was nonetheless a clever decision on behalf of the shop's curators

to pitch climate change stories as a lifestyle subject, not science or politics.

All broccoli and no chocolate makes for a boring and unsustainable diet. All chocolate and no broccoli disrupts the system, too. This metaphor applies to climate change coverage—distressing, moralistic, sombre coverage risks alienating audiences further. Making sure that climate change adaptation can be trendy and exciting prepares us for the long road ahead. In fact, media audiences would like to see some humour in science correspondence (Pinto & Riesch, 2017, as cited in Wicke & Taddicken, 2021: 62), as long as it does not trivialise the matter.

'Light and shade' is a classic paradigm of reporting balance across a newspaper or a website—and it has to be maintained within climate reportage. There are celebrities such as Billie Eilish and Coldplay, among others, who create events with climate awareness in mind. These act as powerful role models for the fans and connect entertainment with climate action.

Billie Eilish, a young American pop sensation, launched Eco-Villages in the locations of her 2022 music tour —fans could fill their water bottles for free, register to vote and be educated on non-profit organisations advocating on climate change, led by women or Black, indigenous or people of colour. During her London concert in 2022, the singer hosted a six-day climate awareness event for the fans—it constituted a conference with climate activists, sustainable fashion brands and documentary screenings (Wang, 2023).

The likes of the highly mediated campaigns like Green Carpet and 30Wears created plenty of newsworthy events to be reported by celebrity press. Livia Firth, an environmental activist and ex-wife of the famous British actor Colin Firth, famously launched the initiative to wear second-hand or upcycled outfits on the red carpet in 2010 (Siegle, 2011). Many celebrities followed her suit, and wearing outfits twice, or reusing old gowns, has become *en vogue* for the 2020s celebrity fashion. Firth went as far as launching the Green Carpet Awards in Milan in 2017 with the help of the renown British environmental journalist Lucy Siegle (Davis, 2022). Since 2023, the Awards are held during the Oscars week in Los Angeles and focus on the celebrities and fashion business that follow eco-friendly practices and avoid pollution, overproduction and waste. #30 Wears was another clever campaign by Livia Firth who asked consumer to estimate whether they could wear the new piece of garment at least 30 times—and if not, resist buying it. This hashtag went viral on social media for some time, with people documenting their reflections on purchasing fashion and reducing impulsive buys.

My own research on fashion media coverage of sustainability (2021) has revealed that much more needs to be done to educate the public on choosing sustainable, long-lasting garments and not falling prey to fast-fashion allure with its poor quality, worker exploitation and short-lasting fabrics. Yet I have also demonstrated that it is an exciting, emerging area of storytelling for professional journalists, bloggers and influencers alike. It calls for a full reconsideration of the fashion prophesying, announcing trends every month or even daily, to encourage consumption. This research on fashion media and sustainability (2021) has revealed 10 patterns of climate coverage—how do fashionista inspire readers and followers to buy more things. Only one pattern identified was related to green consumption—journalists are starting to educate the public on how to buy less and buy better, rent and thrift, reuse and recycle; while the rest of coverage glittered with 'obsession' with the 'must haves' and 'you deserve it' fashion fixes, promoting psychological 'cure' through consumption.

What the work of many fashion media researchers and innovators like Siegle and Firth demonstrates is that there is a need to nourish the new ways of talking about fashion, food, style, living, consumption, architecture and design that does not damage the planet. The climate-friendly lifestyle journalism does not have to be didactic but can be imaginative and creative instead. One example of the novel approaches is the sustainable fashion blogger Aja Barber who recommends non-purchase fashion fixes such as exploring the colour palette of the season and matching the existing items in people's wardrobes to it to achieve new combinations and styles. The opportunities for curious, quirky, intriguing fashion stories are endless.

Same approach applies to food. There is an emerging clarity that 30% of greenhouse gas emissions arise from the food systems—they are generated on all stages of food production and distribution, from deforestation to land use, through to transport and packaging (Crippa et al., 2021; Rippin et al., 2021; Ritchie, 2021). Livestock is responsible for 14% of global greenhouse gas emissions. This is why the scientific consensus is clear—cutting meat is the single most useful change people can implement on an individual level (Dixon, 2022; Westwater, 2021).

Changes to individual diets, restaurant options, festive recipes and beyond—all these areas of coverage create a lot of space for journalistic creativity. There is also a lot of challenge in transforming meat-heavy traditional cuisines of many countries in South America, Europe and the US to the more flexitarian, plant-based options. Consumers need 'considerable assistance', as behavioural psychologists insist. They need to be advised on the exact actions that create a big impact for the climate change reduction—and the media should make this option as easy and attractive as possible. Habit and inertia are strong, as so are cultural ties with meat-heavy recipes and a habit of viewing meat as celebratory meal (think of the Sunday roast in the UK as a once-a-week post-church treat that it used to be).

Other measures in climate-friendly lifestyle reporting can be focussed on reducing food waste, another major source of wasted greenhouse gas emissions. More coverage can be dedicated to offering health-motivated incentives for a considerate, varied and low-meat diet.

Travel journalism that contemplates climate change—especially the high emissions from flying—is starting to suggest local tourism, supported by trains, bicycles and other means of public transport (see Pietrasik, 2019). It draws attention to the walks, staycations and weekends in the vicinity of the places where people live. It propagates the idea that taking a train whenever possible and avoiding flying is a more conscious option. However, this field also sees a spike in some dubious practices that use climate change threat as branding. 'Last Chance tourism' and 'disaster capitalism' (Klein, 2007, as cited in Tegelberg, 2021: 137-138) are some of the terms that criticise the attempts to brand trips to the bleaching coral reefs or the melting Arctic as the apocalypse-framed extravagance.

Which barriers does eco-friendly lifestyle journalism face? It is rooted in capitalist economy of 'more is more' and is often called 'service journalism' (Eide & Knight, 1999). This inherent tie with consumption in an industrial society is the major flaw of service journalism against the need to reduce consumption and consume better. Consumer journalism helps to navigate the complexity of life as a modern global citizen. Craig (2016) aligns with Anthony Giddens' (1991, as cited in Craig, 2016: 127) concept of 'politics of the personal' when individual choices of how to live their lives express individual's 'reflexive project of the self' in the context of overlapping narratives. Standing at the crossroads of globalisation, national media agenda, social media narratives, socioeconomic identity, ethnic and cultural identity, their values and aspirations, a person crafts a version of themselves through things they buy, consume and display.

Consumer lifestyle is not fully at odds with greener living though—'news you can use', as this beat is often called, provides practical instructions and helps to navigate the confusing terrain of pro-climate consumer choices. Directions are essential in building people's self-efficacy in tackling climate change. There is an aesthetic and cultural value to the gatekeeping practice of consumer journalism—reporters and editors seek to draw attention to the ideas, talent and practices that fit the spirit of time, and they act as mediators between cultures (From & Kristensen, 2018). Lifestyle media provide relaxation to the citizens as consumers (From, 2017, as cited in From & Kristensen, 2018: 722). There is an argument that, through service journalism, reporters act as participants in community lives, assistants to their readers who need to make decisions about their money, homes, rights, acceptable practices and habits.

The depiction of ethical consumers in the British newspapers is often grounded in moral superiority or 'cognitive deficiency' (Craig, 2016). Moral superiority may include the criticism of buying clothes on the high street and praise given to the independent 'ethical' brands with prohibitive price tags. 'Cognitive deficiency' angle derives from Craig's (2016) analysis of broadsheet advice sections on green living—the articles are often framed as a dilemma 'how to do the right thing', and the customer is instructed accordingly on such issues as correct recycling, buying from cheaper shops and moving their pensions from those funds associated with fossil fuels.

British tabloids, as opposed to the broadsheet outlets, do not often have specific sections on green lifestyle and tend to publish contradictory materials related to environmentally conscious living. One day a British tabloid The Daily Mail was rooting for the elimination of free plastic bags from the supermarkets, another day it was furious about 'Britons being robbed of their right to buy traditional lightbulbs' (see Craig, 2016).

Lifestyle journalism has a big challenge to overcome—to become inclusive and affordable for various categories of readers. Many solutions offered by current lifestyle reporters come with a hefty price tag or assume a degree of maturity, or a mindset that allows an anti-materialistic existence, with walks in the nature and simple joys instead of consumption on the high street. Organic vegetables, small, crowdfunded clothing brands, local producers or eating out in the places that care about the provenance of food—these practices are still seemingly reserved for metropolitan middle class. Tabloids already show a tendency to politicise these practices and make them part of an ideological 'culture war'. Widening the range of green advice for all categories of earners, people of different ages and backgrounds, cultural habits and customs, will help to keep more readers on board with climate-aware living. Small steps generate attachment to the theme and to the cause—and can help build a sustainable relationship with pro-climate consumer choices.

How High Earners and Social Influencers Can Be Helpful

The individuals that have an influence over society—be it celebrities, wealthy entrepreneurs, notable scientists, Instagram influencers, athletes—have a significant role to play in spreading climate awareness and boosting empowerment. This can be achieved in both traditional media coverage and social media posts and campaigns.

People with higher socioeconomic status (high-SES) are higher emitters. This is true for the developed as much as developing countries—but is much more evident in the Global North. Those in the global top 1% of wealth emit twice as much CO_2 through consumption than the bottom 50% (Kartha et al., 2020, as cited in Nielsen et al., 2021. The exact numbers are 15% and 7% of emissions, respectively).

Influential high earners hold five of roles in a society. They are investors and can prioritise green businesses and sustainable start-ups. They are role models within their social networks—people look up to them for standards and choices. They are participants in the organisations—at either managerial or employee level—and have an impact through discussions, decisions and workplace culture. Then, high-SES people are citizens interested in influencing public policy and societies. Lastly, they are consumers and can make choices in reducing or reshaping their choices and behaviours.

What the influence of high-SES people means for the climate communication is that broadsheet publications, as well as financial media, and the luxury segment publication must incorporate green brands, green initiatives as well as educational but engaging storytelling to engage the high earners with the stories. It is also flattering to remind high-SES people of their power to make an impact and generate a change. Another helpful way to appeal to high-SES for climate cooperation is alluding to 'effective environmentalism', i.e. focussing on the sections of climate issues that can generate tangible results sooner and clearer if the efforts were applied strategically.

'Effective environmentalism' (Boon-Falleur et al., 2022) is the strategy that evaluates which areas of climate emergence have higher chances to bring positive effects and focussing all the power and money on it. It derives from the concept of 'effective altruism' as the approach of supporting charities and causes that are most likely to succeed and generate necessary change (Effective Altruism, 2016). An example of that can be the installation of 200 million insecticide-treated bed nets by Against Malaria Foundation—it is estimated that the initiative has saved around 160,000 lives from malaria (Effective Altruism, 2016).

Involving high earners is crucial for conditional cooperation, i.e. the tendency of people to wait for the influential adopters to lead the way, before they can join in. It can increase the pro-climate behaviour among both high earners and other members of the society who often look up to these individuals for aspiration. Humans have the unique ability to work together to achieve goals, yet our evolutionary development has made us much more prone to pursuing our interests rather than collective ones. To overcome the free-riding effect, humans need to perform three cognitive actions: to detect the social norm that identifies cooperation as the default setting, to demonstrate to the 'herd' that their behaviour fits the norm, and to be clear on what is fair and what is not (Boon-Falleur et al., 2022). Norm detection, reputation management and fairness evaluation are key to sustainable cooperation.

Mid-level Activity and Solutions

Jim Skea, the IPCC (Intergovernmental Panel on Climate Change) Chairman, calls for the pragmatic measures as suggestions for businesses, town planners, landowners—mid-level stakeholders that affect communities (Deutsche Welle, 2023). Targeted advice is the key to ensure that climate mitigation goes well beyond individual choices—groups of people must be empowered to act upon climate change. Infrastructure is key to make sure people can act upon their eco-conscience and climate awareness.

> Individual abstinence is good, but it alone will not bring about the change to the extent it will be necessary', Skea said. 'If we are to live more climate consciously, we need entirely new infrastructure. People will not get on bikes if there are no cycle paths.
> With all these things it's about real people and their real lives, not scientific abstractions. We need to come down a level', Skea insisted. 'There's enough money in the world, the challenge is getting it to flow to the right places. (Deutsche Welle, 2023)

Readers want solutions too (Wicke & Taddicken, 2021). The pivot to the new style of climate journalism in the 2010s was called 'solutions journalism' (McIntyre & Gyldensted, 2018b; Their & Namkoong, 2023) or constructive journalism (McIntyre & Gyldensted, 2018a). Opposed to problem-focussed journalism (highlighting the conflict or social ills), solutions journalism seeks to offer 'treatment recommendation' to the problems identified (Entman, 1993, as cited in Their & Namkoong, 2023: no page). It involves extensive research and fact-checking to ensure that the solutions proposed are at a scale to the problem. From a slightly broader perspective, constructive journalism does not just put the spotlight on the things that are wrong, it presents to the readers the things that are going right and suggests imitating the good practices to heal the ills (Benesch, 1998, as cited in From & Kristensen, 2018). This beat is yet to take shape—there are still debates on how to remain neutral, credible and not overly optimistic about solutions—yet in a nutshell it focuses on the climate-caused issues and deeply researches the existing or possible solutions. It seeks to report not only what is wrong but how to right it.

Some of the caveats of solutions journalism lie in the difficulty of assessing the adequacy of solutions to the crises unfolding. For instance, will the newly developed fertiliser truly resist the droughts and rainfalls and sustain agriculture in the struggling areas? Will the direct carbon-capture technology—still in its nascent phase—emerge as a large-scale affordable solution to reducing the greenhouse gases in the air?..

Moreover, solutions journalism is a helping hand in increasing the accountability of the governments and industry players for eco-damage. This beat of not breaking news, but nonetheless 'hard news' contributes towards practical understanding of the steps that need to be taken to mitigate the climate-related issues (Solutions Journalism Network, an NGO, as cited in Thier & Namkoong, 2023: no page). Although research on the accountability benefits of solutions-oriented climate coverage is lacking (Thier & Namkoong, 2023), it clearly holds potential for working alongside scientists, governments and environmental organisations in identifying reasonable remedies for the climate-induced issues.

Do people appreciate solutions journalism? Various studies suggest that it helps to reduce mistrust in journalism, increases positive sentiments and engagement with the media, including the engagement of under-represented communities, as well as boosts the sense of individual and collective efficacy (Thier & Namkoong, 2023). Yet these ideas have to be taken with a pinch of salt—and need more research—as the empirical proof for theoretical assumptions is still rather limited; theories are not being tested (Lough & McIntyre, 2023). Another constraint lies in the fact that the existing small pool of studies mostly covers the US and Europe.

Curiously, the existing studies uncover that solutions journalism does not necessarily reduce eco-anxiety (Thier & Lin, 2022). Solutions journalism may seem like the progressive way to ignite hope and boost efficacy, but it still features an upsetting message at its core—that does not go unnoticed by the readers. In fact, the emotional response to reading problem-oriented and solutions-oriented journalism was very similar in a 2022 experiment—over half of the audience felt aroused and upset when reading both. There was not any significant sighting of positive emotions associated with solutions journalism (Thier & Lin, 2022). Nonetheless, solutions journalism does come with a promise of efficacy within—it helps to encourage more support for collective climate action (Ibid.).

One of the strong examples of solutions journalism initiatives is Bloomberg Green, a section of the financial outlet Bloomberg that focuses on the green developments in finance, technology, science, culture and society.

"From a Bloomberg perspective the business of climate change and the business of the energy transition struck me as a story that wasn't being told well enough across the news landscape," says John Fraher, Bloomberg's senior executive editor for ESG (Environmental, Social, and Governance) standards and energy—Aisha Majid (2022) reports in Press Gazette. This area of financial developments is a rich field for reportage but is also awash with pledges and promises, as well as greenwashing—the attempts to present something superficial and insignificant as a victory over climate issues. Challenging greenwashing claims, campaigns and brands is also part of solutions journalism, Bloomberg Green assures.

"I actually view that as solutions journalism as well… Part of our role is to educate readers into understanding what is a good solution, a credible and valid solution and what's a phoney solution," says Fraher (as reported by Majid, 2022).

There are similar initiatives within other media outlets—some establish a dedicated section (such as Fixes in The New York Times), others embed solutions-oriented approach across their climate storytelling. There are non-profit organisations and alternative media, as well as influencers that trace solutions-oriented stories.

Solutions journalism existed before climate change—yet with the colossal scale of the problem, more answers-focussed inquiries are desperately needed to keep the public hopeful.

Empowerment

Localisation, stories of people who made a difference and examples of agency from 'people like me' from all around the world can create a feeling of shared achievement. 'Telling local stories, inspiring community-level resistance and transformation and amplifying counternarratives for people who are becoming active citizens' (Painter, 2019: no page) is essential for boosting the sense of empowerment. The IPCC reports call for systemic change—and it is down to the national governments but also to the journalists to break this down into actions that individuals, groups, industries and workplaces can feasibly implement.

The Guardian's campaign 'Keep It in the Ground' in 2015 called for the divestment of funds from fossil fuel-related companies. The media group started with selling its own £800-worth investment in shares, despite their profitability. Similar initiatives and demonstrations followed, with more pressure put on the banks, wealthy individuals, institutions and firms. As a result, in 2020, half of the UK universities joined the pledge to divest from fossil fuels. The mayors of London and New York committed to divest pension funds from fossil fuel assets. This is an empowerment story of a media-generated public pressure that arises from the newsroom conversations and seeks to change the fortunes of fossil fuel companies through market mechanisms.

Micro-credit schemes in Bangladesh, alternative coffee crops in Uganda (Painter, 2019) and innovative iron rust start-ups to generate power in the US (Rathi, 2023)—these are community and society-level stories that need to be highlighted. Realising the amount of work and hope that there is in the world is crucial for empowering citizens to see light and do better themselves.

Empowerment is inseparable from participation. The UNESCO-endorsed framework of HOPE coverage appeals to this, and some of its elements already find use in the newsrooms. HOPE stands for Holistic-future-Oriented-Participatory-Empowering coverage of climate change in the media (Salathong, 2013). It stands on the fundamental belief in explaining the complexity of climate issues—how environment is interlinked with culture and economics (Holistic). Future-oriented means looking for sustainable ways of living and finding better solutions for the existing and upcoming crises. *Participatory* is a big pillar of this media model—citizens need to be informed and equipped with critical skills to be able to contribute to the discussion. Being informed about climate change issues is no longer enough—audience needs to be able to engage with the issues and propose their solutions too (Robie, 2017). It can be accomplished through introducing bottom-up media spaces, such as forum or comment spaces on traditional media websites, or social media channels; establishing dedicated climate-related websites and offline venues. Encouraging voices from the community to raise concerns or doubts, and suggest alternatives is essential to any mitigation and adaptation initiatives, especially involving the marginalised groups. Activating various influential networks to attract people to the participatory communication spaces can be key—in the Pacific Islands, for instance, church and non-governmental organisations have been instrumental in building the awareness of climate issues and adaptation tools available, launching campaigns and educating citizens. *Empowerment* arises from the combination of four—it relies on knowledge and understanding of one's skills and efficacy to motivate for empowered action (Salathong, 2013). Early implementation of this formula by journalists indicates that focussing on one aspect of the paradigm is more realistic than capturing all four (Ibid.).

We are social creatures, and we need social reminders of the positive activity of the others to compare with our own. We need stories of agency from all parts of the world—the troubles of the distant communities need to be reframed in a new light, accentuating humanistic proximity, common lifestyles, stories of nerve and perseverance, stories that trigger empathy and solidarity, and stories that show communities re-organising and curbing the climate problem slowly but steadily, with efficacy and empowerment.

Bibliography

Ahern, J., Galea, S., Resnick, H., & Vlahov, D. (2004). Television images and probable posttraumatic stress disorder after September 11. *Journal of Nervous and Mental Disease, 192*, 217–226.

Al-Rawi, A. (2019). Viral news on social media. *Digital Journalism, 7*(1), 63–79.

Alexander, J. C. (2004). Toward a theory of cultural trauma. *Cultural Trauma and Collective Identity, 76*(4), 620–639.

Ask, K., & Abidin, C. (2018). My life is a mess: Self-deprecating relatability and collective identities in the memification of student issues. *Information, Communication & Society, 21*(6), 834–850.

Ball-Rokeach, S. J. (1985). The origins of individual media-system dependency. *Communication Research, 12*, 485–510.

Bawden, D. (2001). Information overload. *Library & Information Briefings, 92*, 1–15.

Bawden, D., & Robinson, L. (2008). The dark side of information: Overload, anxiety and other paradoxes and pathologies. *Journal of Information Science, 35*(2), 180–191. https://doi.org/10.1177/0165551508095781

Beckett, C., & Deuze, M. (2016). On the role of emotion in the future of journalism. *Social Media+ Society, 2*(3), https://doi.org/10.1177/2056305116662395

Berg, M. (2017). The importance of cultural proximity in the success of Turkish dramas in Qatar. *International Journal of Communication, 11*, 16.

Berger, J., & Milkman, K. L. (2012). What makes online content viral? *Journal of Marketing Research, 49*(2), 192–205.

Berlyne, D. E. (1960). *Conflict, arousal and curiosity.* McGraw-Hill.

Bodas, M., Siman-Tov, M., Peleg, K., & Solomon, Z. (2015). Anxiety-inducing media: The effect of constant news broadcasting on the well-being of Israeli television viewers. *Psychiatry, 78*(3), 265–276.

Bonn Institute (no date). *What is constructive journalism.* https://www.bonn-institute.org/en/what-is-constructive-journalism#the-three-elements-of-constructive-journalism-58480

Boon-Falleur, M., Grandin, A., Baumard, N., & Chevallier, C. (2022). Leveraging social cognition to promote effective climate change mitigation. *Nature Climate Change, 12*(4), 332–338.

Boykoff, M. T., & Boykoff, J. M. (2007). Climate change and journalistic norms: A case-study of US mass-media coverage. *Geoforum, 38*(6), 1190–1204.

Brulle, R. J., & Norgaard, K. M. (2019). Avoiding cultural trauma: Climate change and social inertia. *Environmental Politics.*

Craft, S., & Wanta, W. (2004). Women in the newsroom: Influences of female editors and reporters on the news agenda. *Journalism & Mass Communication Quarterly, 81*(1), 124–138.

Craig, G. (2016). Political participation and pleasure in green lifestyle journalism. *Environmental Communication, 10*(1), 122–141.

Crippa, M., Solazzo, E., Guizzardi, D., Monforti-Ferrario, F., Tubiello, F. N., & Leip, A. J. N. F. (2021). Food systems are responsible for a third of global anthropogenic GHG emissions. *Nature Food, 2*(3), 198–209.

Dai, L., Shen, R., & Zhang, B. (2021). Does the media spotlight burn or spur innovation? *Review of Accounting Studies, 26*, 343–390.

Davis, J. (2022, April 22). Unstitched: Eco-Age's Livia Firth on working together for a greener future. *Harper's Bazaar.* https://www.harpersbazaar.com/uk/fashion/what-to-wear/a39723812/livia-firth-sustainable-fashion/

de Hoog, N., & Verboon, P. (2020). Is the news making us unhappy? The influence of daily news exposure on emotional states. *British Journal of Psychology, 111*(2), 157–173.

Della Porta, D., & Parks, L. (2014). Framing Processes in the Climate Movement: From climate change to climate justice 1. In *Routledge handbook of the climate change movement* (pp. 19–30). Routledge.

Denisova, A. (2021). 'Viral journalism', is it a thing? Adapting quality reporting to shifting social media algorithms and wavering audiences. In *The Routledge Companion to Political Journalism* (pp. 271–278). Routledge.

Denisova, A., (2023). Viral journalism. Strategy, tactics and limitations of the fast spread of content on social media: Case study of the United Kingdom quality publications. *Journalism, 24*(9), 1919–1937.

Deutsche-Welle (2023, July 30). *Don't overstate 1.5 degrees C threat, new IPCC head says.* https://www.dw.com/en/climate-change-do-not-overstate-15-degrees-threat/a-66386523

Dixon, R. (2022, April 28). Climate change: Eating less meat is one of the best ways to reduce your impact on global warming, so why doesn't government policy reflect that? – Dr Richard Dixon. *The Scotsman.* https://www.scotsman.com/news/opinion/columnists/climate-change-eating-less-meat-is-one-of-the-best-ways-to-reduce-your-impact-on-global-warming-so-why-doesnt-government-policy-reflect-that-dr-richard-dixon-3670723

Downs, A. (1972). Up and down with. *Issue-Attention Cycle." The Public Interest, 28,* 51–64.

Effective Altruism (2016). Introduction to Effective Altruism. *The Centre for Effective Altruism.* https://www.effectivealtruism.org/articles/introduction-to-effective-altruism/

Eide, M., & Knight, G. (1999). Public/private service: Service journalism and the problems of everyday life. *European Journal of Communication, 14,* 525–547. https://doi.org/10.1177/0267323199014004004

Elgin, B., Marsh, A., & de Haldevang, M. (2023, March 24). Faulty credits Tarnish Billion-Dollar carbon offset seller. *Bloomberg Green.* https://www.bloomberg.com/news/features/2023-03-24/carbon-offset-seller-s-forest-protection-projects-questioned?leadSource=uverify%20wall

Engesser, S., & Brüggemann, M. (2016). Mapping the minds of the mediators: The cognitive frames of climate journalists from five countries. *Public Understanding of Science, 25*(7), 825–841.

Ereaut, G., & Segnit, N. (2006). *Warm words: How are we telling the climate story and can we tell it better?* (p. 32). Institute for Public Policy Research.

From, U., & Nørgaard, N. (2018). Rethinking Constructive Journalism by Means of Service Journalism. *Journalism Practice, 12*(6), 714–729. https://doi.org/10.1080/17512786.2018.1470475

Futerra, S. (2006). Climate fear v climate hope: Are the UK's national newspapers helping tackle climate change. *Futerra London.* http://www.futerra.co.uk/downloads/Climate_Fear_v_Climate_Hope_Sundays_and_Dailys.pdf

Galtung, J., & Ruge, M. H. (1965). The structure of foreign news. *Journal of Peace Research, 2*(1), 64–91. http://www.jstor.org/stable/423011

Gerbner, G., Gross, L., Morgan, M., & Signorielli, N. (1986). Living with television: The dynamics of the cultivation process. *Perspectives on Media Effects, 1986,* 17–40.

Gifford, R. (2011). The dragons of inaction: Psychological barriers that limit climate change mitigation and adaptation. *American Psychologist, 66*(4), 290.

Harcup, T., & O'Neill, D. (2001). What is news? *Galtung and Ruge Revisited. Journalism Studies, 2*(2), 261–280.

Harcup, T., & O'Neill, D. (2017). What is news? News values revisited (again). *Journalism Studies, 18*(12), 1470–1488.

Havrylets, Y., Tukaiev, S., Rizun, V., & Khylko, M. (2018). State anxiety, mood, and emotional effects of Negative TV news depend on burnout. Preprint. Available at: https://osf.io/preprints/psyarxiv/m3xv2_v1. DOI:10.31234/osf.io/m3xv2

Hayhoe, K. (2021). *Saving us: A climate scientist's case for hope and healing in a divided world*. Simon and Schuster.

Hopke, J. E. (2020). Connecting extreme heat events to climate change: Media coverage of heat waves and wildfires. *Environmental Communication, 14*(4), 492–508.

Ju, H. (2020). Korean TV drama viewership on Netflix: Transcultural affection, romance, and identities. *Journal of International and Intercultural Communication, 13*(1), 32–48.

Keltner, D., & Haidt, J. (2003). Approaching awe, a moral, spiritual, and aesthetic emotion. *Cognition and Emotion, 17*(2), 297–314.

Lachlan, K. A., Spence, P. R., & Nelson, L. D. (2010). Gender differences in negative psychological responses to crisis news: The case of the I-35W collapse. *Communication Research Reports, 27*(1), 38–48.

Lazarus, R. S. (1991). Goal congruent (positive) and problematic emotions. In R. S. Lazarus (Ed.), *Emotion and adaptation*. Oxford University Press.

Limov, B. (2020). Click it, binge it, get hooked: Netflix and the growing US audience for foreign content. *International Journal of Communication, 14*, 20.

Lough, K., & McIntyre, K. (2023). A systematic review of constructive and solutions journalism research. *Journalism, 24*(5), 1069–1088. https://doi.org/10.1177/14648849211044559

Majid, A. (2022, October 27). Solutions journalism: Could it be the antidote to news avoidance? *Press Gazette*. https://pressgazette.co.uk/publishers/digital-journalism/solutions-journalism-news-avoidance/

Maran, D. A., & Begotti, T. (2021). Media exposure to climate change, anxiety, and efficacy beliefs in a sample of Italian university students. *International Journal of Environmental Research and Public Health, 18*(17), 9358.

Mast, C., Huck, S., & Zerfass, A. (2005). Innovation csommunication. *Innovation Journalism, 2*(4), 165.

McIntyre, K., & Gyldensted, C. (2018a). Constructive journalism: An introduction and practical guide for applying positive psychology techniques to news production. *The Journal of Media Innovations, 4*(2), 20–34.

McIntyre, K., & Gyldensted, C. (2018b). Positive psychology as a theoretical foundation for constructive journalism. *Journalism Practice, 12*(6), 662–678.

McNaughton-Cassill, M. (2001). The news media and psychological distress. *Anxiety, Stress and Coping, 14*(2), 193–211. https://doi.org/10.1080/10615800108248354

Meehl, G. A., Karl, T., Easterling, D. R., Changnon, S., Pielke, R., Jr., Changnon, D., Evans, J., Groisman, P. Y., Knutson, T. R., Kunkel, K. E., & Mearns, L. O. (2000). An introduction to trends in extreme weather and climate events: Observations, socioeconomic impacts, terrestrial ecological impacts, and model projections. *Bulletin of the American Meteorological Society, 81*(3), 413–416.

Millan, L. (2023, January 25). Elephant diet choices are helping fight climate change. *Bloomberg Green.* https://www.bloomberg.com/news/articles/2023-01-25/elephant-diet-choices-are-helping-fight-climate-change#xj4y7vzkg

Molek-Kozakowska, K. (2018). Popularity-driven science journalism and climate change: A critical discourse analysis of the unsaid. *Discourse, Context & Media, 21*, 73–81.

Nielsen, K. S., Nicholas, K. A., Creutzig, F., Dietz, T., & Stern, P. C. (2021). The role of high-socioeconomic-status people in locking in or rapidly reducing energy-driven greenhouse gas emissions. *Nature Energy, 6*(11), 1011–1016.

Nolen-Hoeksema, S. (2000). The role of rumination in depressive disorders and mixed anxiety-depressive symptoms. *Journal of Abnormal Psychology, 109*, 504–511.

Painter, J. (2007, June). All doom and gloom? International TV coverage of the April and May 2007 IPCC reports. In *RISJ/ECI conference,* Oxford (p. 26).

Painter, J. (2019). Climate change journalism: Time to adapt. *Environmental Communication, 13*(3), 424–429.

Peters, C. (2011). Emotion aside or emotional side? Crafting an 'experience of involvement' in the news. *Journalism, 12*(3), 297–316.

Phillips, T., Vargas, C., Graham, M., Couch, D., & Gleeson, D. (2023). The victims, villains and heroes of 'panic buying': News media attribution of responsibility for COVID-19 stockpiling. *Journal of Sociology, 59*(2), 580–599.

Pietrasik, A. (2019, October 19). How to explore the world without harming it: Guardian climate pledge 2019. *The Guardian.* https://www.theguardian.com/environment/2019/oct/19/guardian-climate-pledge-2019-air-travel-advice

Piff, P. K., Dietze, P., Feinberg, M., Stancato, D. M., & Keltner, D. (2015). Awe, the small self, and prosocial behavior. *Journal of Personality and Social Psychology, 108*(6), 883.

Porcher, J. E., & de Luca-Noronha, D. (2021). Awe at natural beauty as defeasible evidence for the existence of god. *Manuscrito, 44*, 489–517.

Rathi, A. (2022, November 21) Inside the Billion-Dollar market for climate offsets. *Bloomberg Green.* https://www.bloomberg.com/news/articles/2022-11-21/junk-carbon-offsets-allow-companies-to-claim-they-re-carbon-neutral?leadSource=uverify%20wall

Rathi, A. (2023). *Climate capitalism: Winning the race to zero emissions and solving the crisis of our age.* Greystone Books Ltd.

Renjith, R. (2017). The effect of information overload in digital media news content. *Communication and Media Studies, 6*(1), 73–85.

Rippin, H. L., Cade, J. E., Berrang-Ford, L., Benton, T. G., Hancock, N., & Greenwood, D. C. (2021). Variations in greenhouse gas emissions of individual diets: Associations between the greenhouse gas emissions and nutrient intake in the United Kingdom. *PLoS ONE, 16*(11), e0259418. https://doi.org/10.1371/journal.pone.0259418

Ritchie, H. (2021, November 1). Stop telling kids they'll die from climate change. *Wired.* https://www.wired.co.uk/article/climate-crisis-doom

Rivers, C. (2008). *Selling anxiety: how the news media scare women.* Upne.

Robie, D. (2017). Timely climate media strategy to empower citizens. *Pacific Journalism Review: Te Koakoa, 23*(2), 221–224.

Rosling, H., Rosling, O., & Ronnlund, A. (2018). *Factfulness.* Sceptre.

Salathong, J. (2013). Thai audiences and journalists' responses to a holistic, future-oriented, participatory and empowering (HOPE) model for climate change coverage. *International Journal of Media & Cultural Politics, 9*(1), 71–85.

Schultz, I. (2007). The journalistic gut feeling: Journalistic doxa, news habitus and orthodox news values. *Journalism Practice, 1*(2), 190–207.

Shanahan, M. (2007). *Talking about a revolution: climate change and the media.* International Institute for Environment and Development (IIED), a non-profit research institute, December. https://www.osti.gov/etdeweb/servlets/purl/22059773

Shiota, M. N., Keltner, D., & Mossman, A. (2007). The nature of awe: Elicitors, appraisals, and effects on self-concept. *Cognition and Emotion, 21*(5), 944–963.

Shoemaker, P. J. (1996). Hardwired for news: Using biological and cultural evolution to explain the surveillance function. *Journal of communication.*

Shrimsley, R. (2023, February 1). Conservatives are forced to face the economic case for net zero. *The Financial Times.* https://www.ft.com/content/6ac36f4d-6049-4de0-8e45-d32852b84557

Siegle, L. (2011). *To die for: Is fashion wearing out the world?* Harpers Collins.

Simpson, A., & Tamayo, A. (2020). Real effects of financial reporting and disclosure on innovation. *Accounting and Business Research, 50*(5), 401–421. https://doi.org/10.1080/00014788.2020.1770926

Soroka, S. N. (2006). Good news and bad news: Asymmetric responses to economic information. *The Journal of Politics, 68*(2), 372–385.

Straubhaar, J. (2014). Choosing national TV: Cultural capital, language, and cultural proximity in Brazil. In *The impact of international television* (pp. 77–110). Routledge.

Straubhaar, J. D. (1991). Beyond media imperialism: Asymmetrical interdependence and cultural proximity. *Critical Studies in Mass Communication, 8*, 3959.

Strzałkowski, P. (2023, July 21). *To report on climate change where coal is king, journalists need to focus on solutions*. Reuters Institute at Oxford University. https://reutersinstitute.politics.ox.ac.uk/news/report-climate-change-where-coal-king-journalists-need-focus-solutions

Sunstein, C. R., Bobadilla-Suarez, S., Lazzaro, S. C., & Sharot, T. (2017). How people update beliefs about climate change: Good news and bad news. *Cornell Law Review, 102*(6), 1431–1444.

Tegelberg, M. (2021). Negotiating conflicting temporalities in Canadian Arctic travel journalism. *In Climate Change and Journalism* (pp. 136–151). Routledge.

Tenenboim-Weinblatt, K., & Baden, C. (2018). Journalistic transformation: How source texts are turned into news stories. *Journalism, 19*(4), 481–499.

Thier, K., & Lin, T. (2022). How solutions journalism shapes support for collective climate change adaptation. *Environmental Communication, 16*(8), 1027–1045.

Thier, K., & Namkoong, K. (2023). Identifying major components of solutions-oriented journalism: A review to guide future research. *Journalism Studies*, 1–18.

Thøgersen, J. (2021). Consumer behavior and climate change: Consumers need considerable assistance. *Current Opinion in Behavioral Sciences, 42*, 9–14.

Trepte, S. (2008). *Cultural proximity in TV entertainment: An eight-country study on the relationship of nationality and the evaluation of US prime-time fiction.*

Villi, M., Aharoni, T., Tenenboim-Weinblatt, K., Boczkowski, P. J., Hayashi, K., Mitchelstein, E., Tanaka, A., & Kligler-Vilenchik, N. (2022). Taking a break from news: A five-nation study of news avoidance in the digital era. *Digital Journalism, 10*(1), 148–164.

Wahl-Jorgensen, K. (2013). The strategic ritual of emotionality: A case study of Pulitzer Prize-winning articles. *Journalism, 14*(1), 129–145.

Wang, J. (2023, January 4). Our future: Billie Eilish on climate activism and radical hope. *Vogue*. https://www.vogue.com/article/billie-eilish-climate-activism-january-cover-2022-video

Westwater, H. (2021, August 11). How does eating less meat help the planet? *The Big Issue*. https://www.bigissue.com/news/environment/how-does-eating-less-meat-help-the-planet/

Wicke, N., & Taddicken, M. (2021). I think it's up to the media to raise awareness. Quality expectations of media coverage on climate change from the audience's perspective. *Studies in Communication Sciences, 21*(1), 47–70.

Wilson, K. M. (2000). Drought, debate, and uncertainty: Measuring reporters' knowledge and ignorance about climate change. *Public Understanding of Science, 9*(1), 1.

Wurman, R. S. (1989). *Information anxiety.* Doubleday.

CHAPTER 3

Global South and Global North: Discrepancies in Climate Coverage

GLOBAL SOUTH—HOW IS CLIMATE COVERED IN THE COUNTRIES THAT ARE HIT THE MOST. ISSUES WITH DELOCALISATION, RESOURCES AND LITERACY

The privilege of reading meat-free recipes and carbon-free travel advice is often taken for granted. The storytelling on climate in the Global North seems to have achieved many milestones that the media in developing countries are still trying to reach: the emerging public consensus on the anthropogenic nature of climate crisis; the understanding of the need to act on various levels, from supranational commissions to governments, industries and individuals; more and more science-backed advice is arising for the individuals to instruct them on how to act.

The main issues with climate communication in the Global South derive from the shortage of resources—some linked with the colonial legacy (Boykoff & Roberts, 2007). As a result, many of the less stable and less profitable societies lack the media systems that can respond to the big challenge of climate change accordingly—inform, educate and engage the publics with the issue. The three main challenges of media communication on climate in the Global South in this section are: (1) the lack of interest in science news; (2) the lack of systemic connection of global heating to local events and inability to present a comprehensive big picture of climate harm and ideas for climate action; (3) the need for creative storytelling for the audiences from various backgrounds. It is worth exploring each of them in depth.

1. The recurring trend in climate communication in the countries with less wealthy or diverse media outlets is **the lack of interest in science**. It tends to peak around specific dramatic events, but lays dormant in the routine media environment (see Kakonge, 2011).

In Vietnam, for instance, science journalists struggle to overcome low interest in science news due to the lack of culture of consuming science coverage and due to political control (Tran & Nguyen, 2023). There are three roles of science journalists in the Global South that have to be developed properly in the years to come.

First, science gatekeeping (filtering out important scientific evidence about human impact on natural world and how to mitigate it). Second, science popularisation—making difficult information accessible and, most importantly, engaging. The more ambitious third goal is to empower people to develop critical thinking towards scientific claims—inviting citizens to be engaged collaborators in establishing the limitations and debating the application of scientific results to the policies in a society (Tran & Nguyen, 2023).

2. How to make sure that the drought is related to climate change? Is there a magical tool to prove the connection between the failing crops and the human-induced climate crisis? Do urban areas heat more because it is the summer season, or can the extremes be detected?

Questions like this require avid knowledge, concentration, skill, experience and time—**building a systemic connected discourse on climate is a big ask**. Many Global North newsrooms can afford establishing specific climate or science beats, while organisations with fewer resources in the Global South mostly focus on the daily updates and news incidents (Kakonge, 2011). Problematic business models of the media companies in sub-Saharan Africa, as well as the lack of training to assess science and climate-related information, result in the difficulty in creating a clear analysis of climate science for the audiences. Kakonge (2011) argues for the partnerships with knowledge providers that can train the newsrooms—nonetheless, the barrier to this lies in the costs of those opportunities. Some emerging assistance is provided by the South-to-South initiatives facilitated by the UN that aim at presenting climate journalism as humanised, accessible and relevant to the readers in developing countries (Kakonge, 2011). Without the investment in staff and training, local coverage in many developing countries in sub-Saharan Africa and Asia remains imperilled by the phenomenon that I call *delocalisation*—despite witnessing the extreme weather events literary outside the window, journalists raise their eyes to the international agenda and politics—they miss the opportunity to embed climate change within local contexts.

Okomo-Okello (2009) also defines this shortfall as the issue of quantity, quality and positioning. Not enough articles are published on climate change, researcher argues, and their quality and positioning prevent the reader from amalgamating the effects of warming in a single picture, understanding it as a phenomenon with multiple causes and direct neighbourhood consequences. Kayula (2009) agrees—a review of climate communication in Mozambique, Swaziland and Zambia revealed a lack of coordinated and purposeful communication about climate and therefore failings in raising the awareness of the stakeholders. The countries affected most severely—those of sub-Saharan Africa—are not receiving sufficient engaging, connected and empowering climate coverage (Kakonge, 2011).

3. How do you tell the climate change story to the communities affected the most? Striking them with headlines of more doom and gloom to come is helpful in the immediate reaction and aftermath, yet preparation and activism call for a variety of formats and approaches. To mitigate the lack of engagement with climate coverage, a postcolonial approach is adopted. Many ideas recommended in this book will apply greatly to the Global South, while some will have to be adjusted to the specific context.

The power of storytelling, powerful acts of narrative is undeniable. In India (Uttarakhand in the Himalayan region), women who face the lack of water supplies and shrinking forests have employed storytelling to improve the climate change awareness of their fellow citizens and motivate them for action. They had already been organised in local networks focussed on soil fertility management and crop harvesting (Ravera et al.,

2016). Although situated in the patriarchal male-dominated area, women have been able to organise and adapt through the sharing of seeds and practices and have been more agile in the face of adverse climate-related conditions (Ravera et al., 2016). The inequality of caste, age, gender, background and education affect the way people engage with climate challenges and climate information. The intersectional feminist approach—acknowledging and addressing the overlapping characteristics of individuals and their communities—is helpful to prepare bespoke storytelling for each group. Similar examples of creative, context-aware storytelling are available from other parts of the world—for instance, in sub-Saharan Africa, radio dramas have been identified as a clever and efficient tool to improve the awareness of the HIV/AIDS (Kakonge, 2011).

Another issue is the lack of literacy. While journalists struggle with the head-scratching tasks of translating compound science into accessible formats, many readers—especially in rural communities of sub-Saharan Africa—do not have reading or writing skills, even in their own languages (Kakonge, 2011). To overcome this limitation, more multimedia formats—like radio features, podcasts (Luganda, 2005) or video documentaries—can be introduced to improve the accessibility of storytelling and make the most vulnerable engaged too.

Looking into a specific case study is helpful to understand the many issues—global and regional—that a country or a particular area is facing, and how communication can help. **Africa** has the most youthful population on the planet—and it will stay this way for decades. It also has the world's highest birth rate and scarcity of resources that make it harder for the population to sustain itself. It is predicted that there will be two billion Africans by 2050, and they need agricultural security, robust infrastructure, more energy provision and more telecommunication to be established for the dignified life (Tagbo, 2010). One in three people in sub-Saharan Africa leave in chronic hunger, and running out of foods when crops are hit by drought is a grim commonality (Tagbo, 2010).

Food security is probably the biggest threat that Africa must manage. Having less access to foods due to droughts, floods and sudden changes in the weather, people either have to buy food on very limited financial means, or migrate, or engage in conflicts over more stable land. Water supply is similarly limited—as Lake Chad is now half the size of what it used to be 35 years ago, rainfall is the main source of water. Due to climate change, it becomes hugely unpredictable. All these calamities risk to undermine the progress that African countries have achieved so far—Evelyn Tagbo (2010) argues for the need to convince Global North countries to invest in climate adaptation and at the same time cut the emissions to avoid further harm.

In 2010–2020s, it was estimated that Africa needed around 30 billion dollars *annually* for adaptation to climate change (Ighobor, 2022). Yet by early 2020s, Africa had received around 29 billion dollars instead of 250 billion needed (Savage, 2022). Issues with local governance, lack of skill, infrastructure and data have been cited as the reasons that hold back Western investments in Africa. Despite the UN Secretary General Antonio Guterres's advice to channel fossil fuel taxes into the support for Africa (Ighobor, 2022), developed countries remain cautious about the amounts of funds they send to the continent. Central African countries were particularly perceived as less safe areas for investment due to political turbulence (Savage, 2022).

African Adaptation Acceleration Programme, a joint initiative of the African Development Bank (AFDB) and the Global Center on Adaptation (GCA), is where plenty of international money is flowing to. It promises to tackle data-poor agricultural sector, improve education in sustainable skills for the African youth and develop urban and rural infrastructure in terms of water and energy supply, transport and waste management, creating a robust financial system with tools for financial adaptation and improving access to international funds (AFDB.org, no date).

In the light of these multiple threads of climate threats and weaves of climate adaptation, the local media in Africa are uneven, with the weakness in coverage attributed to funding, political system and trust in journalism. Some of the more developed countries—Nigeria, South Africa and Ghana—feature a rather diverse media system (Tagbo, 2010). The focus on climate change, though, is skewed towards international agenda—the likes of the IPCC reports or COP events, rather than linking local weather happenings with the influence of the climate change.

How much do people on the African continent know about climate change? In 2019, 28% considered themselves literate about the crisis, while a whopping 40% were not familiar with the concept (a survey of 45,000 respondents from 34 African countries—Selormey & Logan, 2019). X (Twitter) users from 54 African countries do not seem to spend much time talking about climate change either—and whey they do so, it does not seem to draw attention to the pressing local issues (Pointer & Matsiko, 2023).

A curious establishment has been detected—people in different areas have each their own definition of what 'climate change' is (Yaro, 2013)—for a fisherman, it is the extreme stormy weather, while for a farmer it is the diminishing crops. While both cases are true and related to climate change, seeing the overall cognition of what climate change is and how it affects most areas of life *locally* in a systemic manner is lacking from the African media coverage (Kayula, 2009; Kakonge, 2011; Okomo-Okello, 2009; Pointer & Matsiko, 2023; Tagbo, 2010). Socioeconomic differences between African countries play a significant role in the variety and depth of coverage. Kenya, South Africa and Nigeria lead on the quantity and quality of climate coverage, while Malawi, Egypt and Uganda feature the lowest number of climate stories in mainstream media (Pointer & Matsiko, 2023).

The focus on immediate disasters coexists with the reporting of the global climate change agenda—largely driven by the high-profile international meetings, UN, COP, release of the IPCC reports, agreements and funds. The peculiar gap between the two—delocalisation—prevents local audiences from engaging with the clear understanding of the issue and preparing better for the next consequences of it. Boykoff and Roberts (2007) cite a curious story—as one of the authors was living in Honduras during the 1998 Mitch hurricane, he witnessed first-hand how the media coverage of the disaster was commanded by the swarms of international correspondents, mostly from the neighbouring US, that ended up framing the issue for the local outlets too. Boykoff calls this asymmetry 'structural disadvantage' (Boykoff & Roberts, 2007: 19), as the immediate effects were immensely illuminated by the international media attention, while preparation to the extreme weather events, as well as subsequent adaptation to the climate-related changes in Honduras, was largely left in the dark.

There are signs that 'a slight shift' is starting to happen, with more coverage devoted to adaptation and mitigation efforts across a range of African media—yet this trend still coexists with the frames of disaster (Pointer & Matsiko, 2023). Journalists in the more stable economies—Kenya, for instance—are often aware of the need to publish more success stories and look into solutions and innovation, less disastrous and alarmist narratives (Osindo, 2014, as cited in Pointer & Matsiko, 2023: no page). Yet the stories focussed on the negative impact prevail.

The positive discovery lies in the fact that—perhaps owing to the limited quantities of climate change coverage—African media have been rather immune to climate deniers (Tagbo, 2010). This means they did not have to deal with that distraction and could focus on the important stories instead. Nevertheless, plenty of climate coverage is coming from the international news agencies (Tagbo, 2010), prompting Gadzekpo (2009, cited in Tagbo, 2010: 34) to criticise local African outlets as 'cheerleaders and amplifiers' of messages from the hegemonic others rather than setting an independent agenda and locally minded conversations. African journalists struggle to adapt complex science to the front-page material, and local angles are not given enough spotlight (Kay & Gaymard, 2021; Tagbo, 2010).

Expanding from the African case study to other regions, the pattern remains roughly the same, with the three dominant issues of media coverage prevailing. Across Thailand, Sri Lanka, Honduras and Jamaica (Harbinson et al., 2006), journalists confessed the lack of knowledge to cover climate science and explain why it matters; the inertia of news values resulted in journalists concentrating on crime and violence; insufficient resources inhibited proper reporting and evaluation of the local impact of the global climate crisis. Even the evident adaptation trends seen in the most affected countries—such as population migrating to urban areas, the launch of recycling and energy efficiency schemes, innovation in agriculture—failed to capture journalists' attention saliently, resulting in patchy coverage. Market competition was often highlighted as another issue—media companies are reluctant to dive into the unknown and adjust their business models and reporting style, unless they are sure that competitors would do the same.

Another issue—not yet documented in the case of many African media systems—is the politicisation of the climate coverage. In some Global South countries, the pattern seen in the Global North—US, UK—is also evident, the pattern for political influence on the quantity and quality of climate information. In Chile, for instance, liberal newspapers publish twice as many stories on climate than the conservative ones and present more variation of framing, not just conflict and government press releases (Dotson et al., 2012).

A 2017 review (Barkmeyer et al., 2017) of climate coverage in 40 + countries (data from 2008) reveals discrepancies in the level of interest to climate change. The more developed countries with democratic governance—Australia, UK—demonstrated a higher number of climate news, while countries with different political systems or less resources for their media—Venezuela, Guatemala, Ecuador, Costa Rica, Russia—were featuring the least amount of content related to climate change. Climate change omission was perhaps more worrisome in some countries than biased climate coverage (Poberezhskaya, 2015). Pakistan and Brazil, surprisingly, were also printing a low number of climate stories, despite being at the receiving end of some of the more dire effects of global warming.

The more prosperous the country is, the more likely are its media to report more on climate change (Barkemeyer et al., 2017). Yet the affluence alone is not enough to ensure prolific storytelling on climate. Effective governance and clear feedback loops play their role too—in the countries where citizens knew how their administrators react to the climate change and how the society can feed back and be heard, the media reported on climate more. Curiously, neither carbon-dependency nor integration in the global economy had much effect on the availability of climate coverage (Barkemeyer et al., 2017). However, Schmidt et al. (2013) correct this by arguing that countries with the stricter carbon emissions target as set by Kyoto Protocol feature more climate stories; however, part of it is explained through the fact that carbon-intensive societies tend to debate the targets and green policies through the media.

Foreign Aid, and How the Media Shape the Agenda in Inter-country Giving

The media coverage plays a significant role in the distribution of foreign aid. It sets the agenda for the developed countries' governors and draws interest of the population towards the places on Earth that need assistance. The way the Western media construct faraway lands matters—it may squeeze or boost the foreign aid spending. For both the donor and the recipient, foreign aid plays part in the formation of national identity and self-perception of its citizens.

It has been documented that whenever there is a spike of Western media interest in a specific disaster or a struggling country, the foreign aid will flow. Politicians monitor national press and influential media outlets; they listen intently to the public opinion on the matter. The more dramatic the storylines are—like that of a woman giving birth on the top of a tree during a flood—the more mass attention and support for funding are likely to follow (Boykoff & Roberts, 2007).

In general, media interest to the distant suffering is an accelerator for foreign aid. The caveat lies in the political inclination of the media outlet reporting on the story. In the UK, for instance, the left-leaning press may be publishing the stories of the struggling lands and populations to contradict the agenda of conservative-led government with its tough stance on migration and reduced empathy capacity—while not necessarily campaigning on the expansion on foreign aid and migration. Yet media attention to the distant suffering and climate effects is more favourable to potential foreign aid generation than media silence.

To prove this, the studies show that one publication in the French newspaper *Le Monde* 'correlates with an additional US$77,000 in aid' to a recipient nation (Van Belle et al., 2004, as cited in Boykoff & Roberts, 2007: 27). Same tendencies have been observed for The New York Times coverage and the corresponding grants from the US donor agencies to the struggling nations. In Japan, the spike in TV and press coverage has led to the increase in loans to those in need of support (Potter & Van Belle, 2004, as cited in Boykoff & Roberts, 2007: 27). Additional geopolitical conditions exist that make the funding more likely—the recipient nation should align in ideology with the donor, and the donor might have been the coloniser of the recipient nation (Hicks et al., 2017, as cited in Boykoff & Roberts, 2007: 28).

Media coverage remains widely asymmetric, with the 'control room' in the Global North and inertia thinking prevailing in the less developed economies. Postcolonial legacy of weak agency, reliance on unsteady business models of the media and a limited access to training and resources for adequate climate coverage mean that local stories remain underdiscussed. They are not fully understood neither by journalists nor by population—and this means the audience that the journalists serve does not have the best information to navigate their lives and engage in adaptation techniques. The 'structural disadvantage' is inseparable from the issue of resources, education, room for experimentation and creativity that only appears when other needs of the newsrooms are met. On the positive side, local understanding of the needs of the population can result in the bespoke projects like radio drama and non-textual modes of communication; these creative initiatives demonstrate the promising modes of interacting with the public on the topic of climate change, increasing accessibility and dissemination of the urgent knowledge.

Overall, the studies on climate change coverage in the Global South feature the issue of **delocalisation that manifests in various forms—the dependence on global climate summits for agenda setting, while local developments triggered by the very same force are largely overlooked**. More science and climate change training should be offered to the reporters from the Global South to make them more prepared and empowered to acknowledge the stories unfolding in front of them and expertly link them to the climate change. Political will is another issue that affects coverage—in the countries with restricted media environments, attributing weather events to climate becomes a challenging endeavour for journalists, if the powers that be deny or resist climate agenda. So far, many countries under scrutiny have been able to escape the populist discourses and manipulative techniques of the fossil fuel lobby. Yet this is a pressure cooker of an issue that academics, governors, media professionals and the civil society should keep an eye on—the Global South is a massive market for growth, which needs energy to achieve the level of life enjoyed by the Global North countries. The carbon promotion permeates the political and media agenda of the developed countries—yet developing countries are ever more vulnerable to these problematic discourses, with fewer resources to debate and resist the fossil fuel lobby in the public domain. The next section explores the shortcomings of media coverage on climate in the carbon-reliant Western societies.

Global North—Progressive Advancements, Precarious Consensus, Political Polarisation and Powerful Lobby

Is it all rosy in the West? As imperfect as the generic nickname for the developed countries is—'Global North'—it nonetheless helps to separate healthier economies from those struggling. Global North usually indicates more mature media systems, with higher variety of quality outlets, and more funds available to support the independent media and the freedom of speech.

On the one hand, big similarities exist—thanks to the globalised nature of a journalist's job, digitalisation of news gathering, globalisation of mass media and the commonplace paid access to the same news wires such as Associated Press, Reuters and Agence France-Presse (Shehata & Hopmann, 2012). There is also a reliance on big agenda-setting English language media doyennes such as the BBC, CNN, Al Jazeera, Sky News, The Times, The Guardian, The New York Times and Washington Post that are largely checked by the media organisations around the world for reference on the prominent events on the day, issues worth reporting about and frames of reportage (McCombs, 2004, as cited in Shehata & Hopmann, 2012: 178).

On the other hand, there are big differences within the 'Global North' bloc—there is Australia, the third largest producer of coal in the world, and there is France that relies on clean nuclear energy; there is the UK that legally committed to net zero by 2050 but gave 100 licenses to new oil and gas exploration in the North Sea in 2023, in a contradictory move. There is Germany led by the green government that nonetheless revived coal plans during the energy crisis in 2022–23. The US is another land of contradictions—since Al Gore and through Donald Trump's reign, top politicians used climate change as a political matter, mobilising voters to highly support or dramatically deny the crisis. This has resulted in very uneven levels of climate action support among Republicans and the Democrats.

What unites the varied countries of the West is the media systems with evident interest in climate change coverage, boasting a variety of it and the ability to localise the events. While media scholars often question how influential traditional media remain in the age of social media and narrowing attention span (Denisova, 2021, 2023; Gilardi et al., 2022), studies show that the more media publications talk about climate, the more the population talks about climate.

In advanced media systems, quality media outlets are more likely to represent objectivity. Yet when the issue is being politicised by the leading political forces, the media reflect this ideological battle. The three countries with the most uneven climate coverage in the Global North are the US, Canada and Australia.

In **Australia**, climate change is often reported through ideological lens. Depending on the government of the day, the country changes its climate outlook dramatically, resulting in what some commentators have called the 20 years of 'climate change wars' (Kelly, 2017, as cited in Lidberg, 2018: 71). Those in support of the conservatives—as well as those possibly lobbied by the fossil fuel industry (Cohen, 2006, as cited in Lidberg, 2018: 71)—debate the anthropogenic nature of the climate change and question the high costs of emissions reduction for the Australian economy (Chubb & Bacon, 2010; Kurz et al., 2010 both cited in Schmidt et al., 2013: 1244). Those in favour of urgent climate action tend to come from the activist groups, few political parties, scientists and even business forces who request a reduction in emissions (Lidberg, 2018). The latter are presenting a curious force to encourage more media coverage of climate change on the continent. According to the Australian journalists, between early and mid-2010s, market players have created a bond with climate groups—their activity generates more climate stories, including those on renewable energy and green initiatives. This bodes well with Rathi's (2023) argument that market forces may be as strong as governments in affecting changes towards climate mitigation and adaptation. For now, the equilibrium in Australian coverage is precarious—there is interest and depth of stories to cover, yet this may change any moment with any new twists on the political scene.

In **Canada**, the reliance on fossil fuels production creates a difficult context for the policymakers and media players. Canada was the only country to withdraw from the Kyoto protocol on emissions reductions after having ratified it (Stoddart et al., 2016). The country is criticised for 'climate delayism' for obstructing rapid change and adhering to 'bridge' solutions like focussing on natural gas while drifting away from coal (Janzwood & Millar, 2022). Canadian national newspapers devote a substantial amount of coverage to climate change but mostly report on policy and governmental responsibility rather than civil society or cultural shift (Stoddart et al., 2016). Little spotlight is given to the activity of environmental groups, publics and their opinion, and bottom-up adaptation initiatives, indicating a pivot to top-down national technocratic solutions.

The **United States of America** is in a paradoxical position—it has been documented by research that many US-based media set the agenda for much of the global climate coverage. The respected transnational mavens–CNN, The New York Times, Washington Post—are quality media titles that solidify climate certainty and deliver a wide range of stories, from the high-level policy development to the low-level adaptation activity on the ground (Shehata & Hopmann, 2012). The US discourse is dynamic and ever-evolving—the 2010s saw the need to encourage the urgency of climate awareness among readers and stakeholders, while mid-2010-2020s demonstrate the need for clearer coverage on the types of high-, mid- and person-level action for both mitigation and adaptation. Adaptation storytelling had been particularly absent until recently (Ford & King, 2015).

Yet the other side of the coin in the US media system is the deep polarisation around climate issues. For a few decades, the fake 'balance' of views on climate change permeated the more popular or conservative-leaning media outlets in America. Climate sceptics had been enjoying the spotlight as much as climate scientists. The turning point only happened in 2005–07 (Feldman et al., 2017). It is only for the last two decades that the influential national media stopped or reduced invitations to climate deniers.

Nevertheless, the coverage even from the most respected media titles in the world is not heterogeneous. For instance, even some of the media considered elite and objective may prioritise negativity and uncertainty frames thus decreasing the sense of efficacy within the audience. Wall Street Journal was identified as one of these titles—in a recent study, it was revealed that it featured a strong focus on conflict frame and the negative economic outcomes of the mitigation efforts; the morality frame that suggests the responsibility for climate action was not salient (Ibid.). Among the popular media, Fox News cable network has been featuring climate deniers and seeding doubt for a long while. Overall, the most influential media in the US—The New York Times, Washington Post, USA Today, Fox News—all lack the middle layer of coverage, the efficacy-boosting stories and suggestions to the collectives and individuals on the course of climate action (Ibid.). When most attention is focussed on the government action, policy, conflicts among political viewpoints on strategies of mitigation and adaptation, citizens may feel sceptical or even left out—which is dangerous and may lead to withdrawal from climate news, and any interest in taking action (Ibid.).

In **Europe**, the increasing heatwaves and green policy initiatives by the bloc and by individual governments are gaining attention in the press. In most European media—e.g. France, Germany, Sweden, Italy—the consensus has emerged over the anthropogenic nature of climate crisis (Brossard et al., 2014, as cited in Pasquaré & Oppizzi, 2012: 153; Shehata & Hopmann, 2012). Despite the growing amount of coverage of climate change, the interest and eagerness to act upon them are not high.

The Mediterranean Basin is a hub of heatwaves with the countries within its vicinity experiencing not only intense heat but also grappling with more severe droughts, storms and extreme weather conditions. Nonetheless, the media storytelling in Southern Europe tends to be state- and institution-focussed (Areia et al., 2019). The lack of science-based coverage and civil society stories makes the discourse somewhat isolated from the efficacy of ordinary citizens. Even when the heatwaves strike, the media often report weather separately from climate. In 2023, for instance, during the July heatwave in Italy, many newspapers used deadly metaphors to describe the intensity of heat in some stories in the middle of the newspapers/websites. However, the front pages were dominated by another '*bufera*' (meaning 'storm')—the political debate to chase tax evaders, generated by the populist deputy prime minister. This asymmetric attention to political whereabouts—at the expense of community and industry attention towards climate change impacts—is similar to the shortcomings of the best of the US press. It creates a disconnected narrative where people wait for action from their elected elites while growing cynical of those at the same time. They end up noninterfering in any activity related to climate change mitigation in the meantime.

Italy is a crystallised case study of the climate coverage trajectory in Europe over years. Italian press used to be divided along the ideological lines until the 2010s—e.g. a left-leaning national newspaper was reporting scientific certainty on the man-made nature of climate change, while the right-leaning press gave more room to the 'negationists' (Pasquaré & Oppizzi, 2012). Inglisa (2008) adds to this that Italian television tended to confuse weather and climate stories—although the research in mid-2010s demonstrates a significant shift towards reporting climate change as indisputable (Beltrame et al., 2017). Television still plays a significant role in informing the citizens about climate news (around 60–70% of Italian respondents in mid-2010s), with one-third of population relying on the Internet and a quarter checking out newspapers for climate stories (Ibid.).

The shortcoming of Italian media approach is that the citizens don't feel the urgency to address the issue—and they lack practical instructions on how to act. People are widely aware of the troubles caused by the climate change—often through witnessing the extreme weather events or consuming the information on them through traditional media—but they remain less informed on the causes and do not hear many scientific voices in the coverage (Ibid.). Curiously, Italians trust television news more than scientists when it comes to the global warming (42% against 30%)—the figure higher than the European average (Ibid.).

'Global, not yet local' is the theme of the narrative in **the Baltic area**. The pattern is similar to the Mediterranean countries—lots of coverage is driven by international agenda and conferences, with minimal domestication of the issues and mitigation initiatives (Kleinberga, 2022).

Perhaps a more promising level of localisation is brewing in specialist media across Europe. Although research is limited, they are pioneering a more community-friendly approach to climate coverage. Professional publications are paying close attention to the particular application of scientific knowledge—e.g. farming magazines are closely monitoring the changes for agriculture, its contribution to the emissions, climate policy's effects on agriculture and adaptation techniques (Asplund et al., 2013).

China is among the top-three global greenhouse gas emitters. It is a powerful economy and an influential political player, making it hard to classify as Global North or Global South. Given the dual affiliation of the Chinese economy—its reliance on mass production of goods for export, and huge investment in the green energy race (Rathi, 2023)—China appears as both a formidable polluter and a pioneering green advocate. The media coverage of climate change is therefore intriguing: Chinese press presents the crisis as a global affair, as an issue that the global community needs to address (Song et al., 2022). The Chinese media system remains under scrutiny of the party that expects its journalists to preserve a certain ideological unity (Duan & Takahashi, 2017, as cited in Song et al., 2022: 2210). So far, Chinese media coverage of the climate issues has been largely defensive of the Chinese official policy, with the curious detail that it has also featured a lot of translated material from the international climate conferences (Tolan, 2007). This shows that even China with its seemingly insulated media system cannot escape the global agenda setters on climate, be it intergovernmental organisations or global media. Another consequence of a unique geopolitical position of China is that its media report the cooperation with the West as much as the South-to-South initiatives, including bilateral programmes with African and Caribbean countries (Song et al., 2022). The consensus over the fact that climate change is happening, and all departments and regions should work towards addressing it, is another evident pattern in Chinese climate storytelling (Ibid.). The shortcoming of this 'unified' voice is that the media coverage in China prioritises the positive news pattern, highlighting achievement and progress, and paying less attention to the critical issues and the plurality of voices (Tolan, 2007).

The Most Prominent Five Frames of Climate Coverage in the Global North

Framing is a powerful conceptual tool that helps analyse media messages along four lines of inquiry: problem, causes, treatment recommendations, and moral evaluation (Entman, 1993). It thus helps to show how the media identify the causes and consequences of the climate-related events, where they attribute blame and whether and where they seek solutions. Not every media piece contains all four elements—most likely, in quality newspapers and websites the 'moral' aspect is to be omitted as it comes with judgement (Wessler et al., 2016). The only tint of 'moral' evaluation can be seen in the singling out of particular countries with higher emissions counts (Ibid.). Other three elements of framing are crucial to understand as they affect the public comprehension of climate change causes, issues and adaptation and mitigation strategy.

The prevailing five frames of climate coverage in Global North up to date have been responsibility, conflict, (economic) consequences, human interest and morality (Boykoff & Roberts, 2007; Semetko & Valkenburg, 2000).

Responsibility frame tends to be on the rise during the times of international conferences such as those like COP or UN summits. It is these intercontinental forums that feature the conversations on high emitters, 'loss and damage' funds and the unequal distribution of emissions across countries. The media report accordingly—across Global North and Global South media units, the high-profile international climate forums generate plentiful headlines on responsibility (Wessler et al., 2016).

In a study that focussed on the media coverage of the UN climate conference across five democracies (Wessler et al., 2016)—the US, Germany, Brazil, India and South Africa—lots of similarities were found. The prevailing frames in the national newspapers across five democracies were 'global warming victims' (around 30% of the sample, stories ranging from people to habitats and wildlife), the civil society demands (from policy interventions to activism), the political negotiations and the sustainable energy frame (Ibid.). It is important to acknowledge that these frames were relevant in the coverage of the specific climate policy events—the UN climate conferences across 2010–13. The nod to energy transition and climate justice tends to get more pronounced during those gatherings.

Conflict may refer to the differing opinions on the anthropogenic nature of climate change—and this used to be discernible in the US coverage until the early 2000s (Boykoff & Boykoff, 2007), yet with the awful damage brought about by the likes of Hurricane Katrina and other weather events, the consensus seems to have shifted towards climate awareness (Boykoff, 2007). In most of Europe, 'conflict[1]' of opinions has never been the mainstream frame and decreased further in mid-2000s, giving way to more unity over the understanding of climate change causes. However, there is a concern that the new 'conflict' frame is emerging from the debates on how to curb the emissions and keep economies stable. There is a tension between the governments that introduce green initiatives or levy and the companies that have to pay the price to adhere to those; many of these policies require sacrifices or investments in the short run but ensure long-term sustainability benefits. The strain over varying understanding of the short-term plan and believing in the long-term goals is acute across US (Grundmann & Scott, 2014, as cited in Song et al., 2022: 2210). In 2023, public opinion polls were showing that most US citizens did not understand the scale and the benefits of the ambitious Inflation Reduction Act by President Biden that poured billions of dollars into greening the economy. The transition from fossil fuels to green energy may be less painful if the media justify it through careful fact-checking, human stories explainers, not emphasis on political conflict.

Consequences is the most used frame—it bodes well with the traditional news values (see Chapter 2 on climate-adjusted news values). The events that are negative and affect people in dramatic ways make headline material—especially if they happen in the geographical or cultural proximity to the readers. The 'consequences' frames can be also defined as 'disaster', 'issues' and 'bad news'. While it provides an immediate dramatic effect and grabs attention, it is not maintainable to drive sustained attention towards the issue. With too much fear and doomism wrapping up the storytelling, it serves as a deterrent to the interest in climate coverage,

---

[1] Sadly, in the years to come the 'conflict' frame may rise to the prominence again—but for different reasons. Conflict over the diminishing resources may lead to fights over water, land, energy—although climate change and security scholars (see Meierding, 2015) argue that the cooperation scenarios should be discussed in the media and policy circles too, as an alternative paradigm and frame of representing the struggling communities.

when not balanced out with the positive stories of action (see Chapter 2 for the suggestions on 'frequency' reporting approach instead).

Adaptation frame may be part of the consequences frame—yet it tends to be widely under-developed. When the US and Canadian press, for instance, cover adaptation scenarios, many outlets seem to be enchanted by techno-engineering aspirations. They overlook the 'soft' adaptation processes that include local practices and resilience (Ford & King, 2015).

Human interest is the frame that brings the issue close to home. This can entail profiling of the individuals that are at the forefront of climate damage, or looking at the specific localised consequences of the global warming. By 2010, French and Dutch media were not using personal human stories much (Dirikx & Gelders, 2010) giving more space to the economic, area-wide consequences and more general scenarios of the catastrophe. However, in the 2020s, the human-interest frame is becoming more saturated—the examples are found during the heatwave of Summer 2023 across Europe. In the UK, Spain, Austria, the Netherlands and Norway, the extreme temperatures made front pages of the national newspapers (Binda et al., 2023). Some of them focussed on a peculiar detail—like the Greek Acropolis being closed for tourists due to the heat, or people evacuated from the resorts of Tenerife due to wildfires. These media stories used references to the globally recognisable symbols, such as the temple of ancient Greek civilisation, or the trouble at a popular tourist destination, to create relatability that made the heatwave's scale ever more present and tangible for the readers.

Morality frame is hard to execute by an objective journalist—the rare way to insert religious or moral judgements into a story is quoting opinion leaders (Dirikx & Gelders, 2010). Research of the media coverage of climate does not feature morality frame much (Ibid.). Nonetheless, the morality frame can be found in the left-leaning publications in Europe that give more space to the coverage of degrowth argumentation, loss and damage funds, critique of excessive consumption in the West.

Overall, the discrepancy of climate change coverage in Global North and Global South is vivid and closely related to historical, economic and literacy reasons. Lack of available training and resources in many newsrooms in the developing countries prevents them from exploring the issues lying in the vicinity and results in the reliance on hegemonic agenda. Plenty of publications in the Global South shape their climate coverage in line with the influential international events like COP summits or the release of new IPCC reports. They often struggle with generating agenda that derives from the local context—yet when they do so, it is the more affluent countries with the more developed media systems that lead on the quantity and quality of the stories in the Global South.

There is a growing tendency to experiment with the creative media formats that are utilised to cater to the needs and capacity of specific audiences. Some examples of engaging and informative storytelling on climate change include radio drama and non-textual stories on climate change aimed at the people in remote and rural areas. In the Global North, the leading patterns of climate storytelling include the increased variety of stories aimed at many levels of initiative and action. Some countries show a tendency to prioritise the actions and words of the governments, thus neglecting the agency of the collectives and individuals. In other contexts, the issue of political polarisation on climate change persists—not as much in the shape of denying the anthropogenic nature of the issue but debating over the curbs and incentives used to mitigate its effects.

Framing of the climate-related events remains an important tool that affects agenda-setting and available discourses on the matter. So far, the prominence of conflict and consequences overshadows the frames of responsibility, morality and human interest. These underutilised frames are recommended for further exploration by the leading newsrooms, which have the power to set the tone for the coverage globally and inform the agenda not just on the topics but the production values of storytelling.

Bibliography

AFDB.org (no date). *Africa adaptation acceleration program.* https://www.afdb.org/en/topics-and-sectors/initiatives-partnerships/africa-adaptation-acceleration-program

Areia, N. P., Intrigliolo, D., Tavares, A., Mendes, J. M., & Sequeira, M. D. (2019). The role of media between expert and lay knowledge: A study of Iberian media coverage on climate change. *Science of the Total Environment, 682,* 291–300.

Asplund, T., Hjerpe, M., & Wibeck, V. (2013). Framings and coverage of climate change in Swedish specialized farming magazines. *Climatic Change, 117,* 197–209.

Barkemeyer, R., Figge, F., Hoepner, A., Holt, D., Kraak, J. M., & Yu, P. S. (2017). Media coverage of climate change: An international comparison. *Environment and Planning c: Politics and Space, 35*(6), 1029–1054.

Beltrame, L., Bucchi, M., & Loner, E. (2017). Climate change communication in Italy. In *Oxford Research Encyclopedia of Climate Science* (pp. 1–35). Oxford University Press.

Binda, R., Chandrasekhar, A., Dunne, D., Lempriere, M., Quiroz, Y., Song, W., Tandon, A., Uwaifo, E., & Viglione, G. (2023. July 18). Media reaction: Extreme weather hits world's seven continents in July 2023. *Carbon Brief*. https://www.carbonbrief.org/media-reaction-extreme-weather-hits-worlds-seven-continents-in-july-2023/

Boykoff, M. T., & Roberts, J. T. (2007). Media coverage of climate change: Current trends, strengths, weaknesses. *Human Development Report, 2008*(3), 1–53.

Denisova, A. (2021). 'Viral journalism', is it a thing? Adapting quality reporting to shifting social media algorithms and wavering audiences. In *The Routledge Companion to Political Journalism* (pp. 271–278). Routledge.

Denisova, A., (2023). Viral journalism. Strategy, tactics and limitations of the fast spread of content on social media: Case study of the United Kingdom quality publications. *Journalism, 24*(9), 1919–1937.

Dirikx, A., & Gelders, D. (2010). To frame is to explain: A deductive frame-analysis of Dutch and French climate change coverage during the annual UN Conferences of the Parties. *Public Understanding of Science, 19*(6), 732–742.

Dotson, D. M., Jacobson, S. K., Kaid, L. L., & Carlton, J. S. (2012). Media coverage of climate change in Chile: A content analysis of conservative and liberal newspapers. *Environmental Communication: A Journal of Nature and Culture, 6*(1), 64–81.

Entman, R. M. (1993). Framing: Toward clarification of a fractured paradigm. *Journal of Communication, 43*(4), 51–58.

Feldman, L., Hart, P. S., & Milosevic, T. (2017). Polarizing news? Representations of threat and efficacy in leading US newspapers' coverage of climate change. *Public Understanding of Science, 26*(4), 481–497.

Ford, J. D., & King, D. (2015). Coverage and framing of climate change adaptation in the media: A review of influential North American newspapers during 1993–2013. *Environmental Science & Policy, 48*, 137–146.

Gilardi, F., Gessler, T., Kubli, M., & Müller, S. (2022). Social media and political agenda setting. *Political Communication, 39*(1), 39–60.

Harbinson, R., Mugara, R., & Chawla, A. (2006). *Whatever the weather: Media attitudes to reporting climate change, project report*. Panos Publications.

Ighobor, K. (2022, November 8). Millions pledged at Africa adaptation acceleration event. UN.org, *Africa Renewal magazine*. https://www.un.org/africarenewal/magazine/november-2022/millions-pledged-africa-adaptation-acceleration-event

Inglisa, M. (2008). The representation of climate change in the Italian media: research report: the analysis of daily national newspapers and TV bulletins of

Rai and Mediaset. [*La rappresentazione dei cambiamenti climatici nei media italiani: rapporto di ricerca: analisi dei principali quotidiani nazionali e dei telegiornali di prima serata Rai e Mediaset*]. Ibis.

Janzwood, J., & Millar, H. (2022, September 11). A bridge to nowhere: Natural gas will not lead Canada to a sustainable energy future. *The Conversation*. https://theconversation.com/a-bridge-to-nowhere-natural-gas-will-not-lead-canada-to-a-sustainable-energy-future-176734

Kakonge, J. O. (2011). The role of media in the climate change debate in developing countries. *Global Policy Essay*, 2.

Kay, N., & Gaymard, S. (2021). Climate change in the Cameroonian press: An analysis of its representations. *Public Understanding of Science, 30*(4), 417–433. https://doi.org/10.1177/0963662520976013

Kayula, F. (2009). Media coverage of climate change in Southern Africa: Lessons from Mozambique, Swaziland, and Zambia. *Panos Institute of Southern Africa*. http://www.fanrpan.org/documents/00738/

Kleinberga, V. (2022). Global, not yet local: Media coverage of climate change and environment related challenges in Latvia. *Information & Media, 93*, 8–27.

Lidberg, J. (2018). Australian media coverage of two pivotal climate change summits: A comparative study between COP15 and COP21. *Pacific Journalism Review, 24*(1), 70–86.

Luganda, P. (2005). *Communication critical in mitigating climate change in Africa*. Open Meeting of the International Human Dimensions Programme, Bonn, Germany.

Meierding, E. (2015). Disconnecting climate change from conflict: a methodological proposal. In *Reframing Climate Change* (pp. 52–66). Routledge.

Okomo-Okello F. (2009). *Media coverage of climate change in East Africa: Importance, challenges, opportunities and needs*. http://ej.msu.edu/eastafrica/powerpoint/okello_Media_Coverage.ppt

Pasquaré, F. A., & Oppizzi, P. (2012). How do the media affect public perception of climate change and geohazards? An Italian case study. *Global and Planetary Change, 90*, 152–157.

Poberezhskaya, M. (2015). Media coverage of climate change in Russia: Governmental bias and climate silence. *Public Understanding of Science, 24*(1), 96–111.

Pointer, R., & Matsiko, S. (2023). How are Africans talking about climate change and who is doing the talking? *Journal of African Media Studies, 15*, Issue Shifting African Narratives, Jun 2023, 247–271.

Rathi, A. (2023). *Climate capitalism: Winning the race to zero emissions and solving the crisis of our age*. Greystone Books Ltd.

Ravera, F., Martín-López, B., Pascual, U., & Drucker, A. (2016). The diversity of gendered adaptation strategies to climate change of Indian farmers: A feminist intersectional approach. *Ambio, 45*(Suppl 3), 335–351. https://doi.org/10.1007/s13280-016-0833-2

Savage, R. (2022, April 11). Africa getting just 12% of financing needed to adapt to climate change -report. *Reuters*. https://www.reuters.com/world/africa/africa-getting-just-12-financing-needed-adapt-climate-change-report-2022-08-11/

Schmidt, A., Ivanova, A., & Schäfer, M. S. (2013). Media attention for climate change around the world: A comparative analysis of newspaper coverage in 27 countries. *Global Environmental Change, 23*(5), 1233–1248.

Selormey, E. E., & Logan, C. (2019, September 23). African nations are among those most vulnerable to climate change: A new survey suggests they are also the least prepared. *Afrobarometer*. https://afrobarometer.org/blogs/african-nations-are-among-those-most-vulnerable-climate-change-new-survey-suggests-they-are. Accessed 22 February 2022.

Semetko, H. A., & Valkenburg, P. M. (2000). Framing European politics: A content analysis of press and television news. *Journal of Communication, 50*(2), 93–109.

Shehata, A., & Hopmann, D. N. (2012). Framing climate change: A study of US and Swedish press coverage of global warming. *Journalism Studies, 13*(2), 175–192.

Song, Y., Huang, Z., Schuldt, J. P., & Yuan, Y. C. (2022). National prisms of a global phenomenon: A comparative study of press coverage of climate change in the US. UK and China. *Journalism, 23*(10), 2208–2229.

Stoddart, M. C., Haluza-DeLay, R., & Tindall, D. B. (2016). Canadian news media coverage of climate change: Historical trajectories, dominant frames, and international comparisons. *Society & Natural Resources, 29*(2), 218–232.

Tagbo, E. (2010). *Media coverage of climate change in Africa*. Reuters Institute for the Study of Journalism.

Tolan, S. (2007). *Coverage of climate change in Chinese media*. https://core.ac.uk/download/pdf/6248686.pdf. Accessed 10 January 2021.

Tran, M., & Nguyen, A. (2023). Besieged from all sides: Impediments to science journalism in a developing country and their global implications. *Journal of Science Communication, 22*(4), A04.

Wessler, H., Wozniak, A., Hofer, L., & Lück, J. (2016). Global multimodal news frames on climate change: A comparison of five democracies around the world. *The International Journal of Press/politics, 21*(4), 423–445.

Yaro, J. A. (2013). The perception of and adaptation to climate variability/change in Ghana by small-scale and commercial farmers. *Regional Environmental Change, 13*, 1259–1272.

CHAPTER 4

The Many Faces of Greenwashing

How the Coffee Takes the Worry Away, or a Tale of Coffee-Washing

A cup of morning coffee can power you through the tasks of the day. Why not send the used coffee waste to power somethings else—like city buses? In 2017, dozens of headlines claimed that biofuel made from coffee waste is the new breakthrough to charge the iconic red London buses. BBC (2017) cited 'technology firm *bio-bean*' in their article, Evening Standard (Morrison, 2017) decisively declared that coffee-fuel 'will be used to help power some of London's buses from today'. The story made headlines across the world—from The New York Times to Architectural Digest and Conde Nast Traveller.

While all the eyes were on the story, not many journalists spotted a sensitive nuance of this innovation. The news piece in The Daily Telegraph only featured the peculiar detail in the sixth paragraph of their piece—Shell, the fossil fuel giant, was sponsoring the research. BBC did not mention it at all.

Four years prior, a start-up called *bio-bean* won Shell's entrepreneur of the year competition and received £5,000 for research. It later entered Shell Springboard programme for small low-carbon enterprises. In 2017, Shell poured money and resources into an expansive public relations campaign on coffee beans and London buses—three communication

© The Author(s), under exclusive license to Springer Nature
Switzerland AG 2025
A. Denisova, *Effective Climate Communication*,
https://doi.org/10.1007/978-3-031-67340-5_4

firms were hired; it encompassed digital and outdoor advertising and resulted in the headlines all over the globe (Tan, 2017).

Why not laud the efforts to decarbonise London buses? The feebleness of the Shell and *bio-bean* initiative was that… only one bus was powered by the coffee-derived biodiesel. This little but not insignificant detail skipped the scrutiny of many media publications—the BBC article that failed to discuss Shell's involvement with the project did mention that the 'coffee-oil' produced up to date will be enough to power a bus for a year. It did not highlight enough that just one bus could benefit from the vaunted campaign. The Evening Standard run the fact of powering one bus in the third paragraph—yet both news stories casually claimed the plural 'buses' throughout their coverage—announcing that coffee-diesel will power 'some' London buses.

This practice of claiming green credentials but not changing the core nature of the carbon-heavy business is known as greenwashing. Another definition of this phenomenon is 'a gap between symbolic and substantial actions' (Siano et al., 2017: 27). The term was coined in the 1980s by the American ecologist Jay Westerveld to criticise the communication that praises small eco-achievements of a company while glossing over major issues of the same company that generate ecological damage (Romero, 2008).

While factually—mostly—correct, the ambitious story about turning Londoners' beloved drink into fuel for double-deckers does not solve anyone's climate issues. Neither does it justify the actions of its funder—the emitter of 1,232,000,000 tons of CO_2 (Bousso & Nasralla, 2023). Not many mentions can be found of *bio-bean*'s efforts to expand the experiment to other buses in London. In 2020, CNN (Thin & Lewis, 2020) reported that the London bus project was not 'commercially viable', and *bio-bean* 'has shifted its focus to solid fuels for household and industrial use'.

Greenwashing takes many forms—from the deceptive or untrue ecological claims made by the companies, industries and governments in their public communication, to the promotion of the organisation's social status as sustainable—all while not engaging with significant changes to the production and business models that could help cut the environmental impact (Nemes et al., 2022). The clear-cut definition of 'greenwashing' is still in the making (ibid.), thus making it harder for regulators to curb dishonest practices. The control over greenwashing is uneven between countries, with some taking a harsher approach with fines and prison terms, and others engaging in monitoring practices and nudges towards green transparency. It is also often unclear which parts of business communication—corporate strategy, corporate reports, stakeholder-oriented communication, or the public-facing public relations and advertising—should fall under direct scrutiny.

The European Union launched the Unfair Commercial Practices Directive in 2005 that monitors greenwashing in business-to-consumer communication, while in Canada in 2008 the advertising and competition authorities ensured that false eco-claims are punishable by fines, product confiscation and imprisonment. Canada, Australia and France implemented a similarly strict regulation (Delmas & Burbano, 2011).

In the UK, the financial authority prepared the Guidance on Green Claims in 2022. Also in the UK, the Advertising Standards Association drafted the—non-legally-binding—code for responsible environmental advertising. The lines are extremely blurred, with deception or 'red herring' techniques being abundant among commercial companies that are trying to 'green' their reputation. Furthermore, as non-governmental and non-profit organisations are not covered by the codes and directives, the guidelines and controls often do not apply to 'non-advertisements' (certificates, partnerships, pledges) (Nemes et al., 2022). As if detecting greenwashing was not challenging enough (Delmas & Burbano, 2011), these additional loopholes make accountability an effortful and complicated endeavour.

In an effort to be as creative in defining greenwashing in its multiple chameleon forms as the firms pursuing this behaviour are, academics have turned to rather colourful terminology. Tiffany Gallicano (2011) proposes seven themes of greenwashing: skeleton in the closet; the right hand isn't talking to the left hand; magic tricks; larger than life; may I have the definition please?; law and order; and truth and fiction. Most of these techniques refer to the diversion of attention from the pollution, poor labour conditions or other unsustainable impact of the core business.

1. 'Skeleton in the closet' is rather self-explanatory—a 'dirty' company adds a layer of virtue-signalling and eco-pledges to divert attention from its core business.
2. 'The right hand isn't talking to the left hand' refers to the eco-claims that may be based on some aspects of the product but not consider its full lifecycle (e.g. a business may claim that a pair of jeans made of cotton is easier to recycle than a polyester garment.

Yet in spotlighting the ease of recycling, it obscures how resource-heavy is the production of jeans—with one pair requiring 15,000 litres of water to make (Denisova, 2021).
3. 'Magic tricks' refers to big assertions that are hard to prove. It can also include irrelevance—a company may highlight its virtues that have little to do with the real environmental impact of the brand (Delmas & Burbano, 2011).
4. 'Larger than life' refers to the exaggerated environmental claims—like a certain bus powered by a certain beverage.
5. 'May I have the definition please?' distinguishes declarations that use pseudo-scientific language or are vaguely defined. 'Eco-friendly', 'biodegradable' and 'all-natural' are common examples of a vague promise that means little, if left unscrutinised (Gillespie, 2008); 'sustainable' is another mean-all-mean-nothing term (Denisova, 2021).
6. 'Law and order' nods to the practice of claiming credit for the 'greening' initiatives that are mandated by the law anyway; it also covers the lobbying efforts that are contrary to the 'green' pledges. An example of this is General Electric's 'Ecoimagination' campaign that put the company's ecological efforts in the spotlight—all while this major American energy provider company was simultaneously lobbying against new clean air regulation (Delmas & Burbano, 2011).
7. 'Truth and fiction' indicates the false endorsements or certifications that the company pretends to have. Lyon and Montgomery (2015) add to this the co-opted partnerships with NGOs and certification providers.

Challenging greenwashing in communication is hard. As I write in the chapter on news values for climate change, the newsrooms need training and resources to be able to debunk the vigorously constructed claims and fact-check the vaguely phrased pledges. Journalists need training and resources to investigate deeper issues such as business models of companies, the science behind the lifecycle of a product, the validity and worthiness of certifications and accreditations. This is a mammoth, but not impossible task.

Paradoxically, those companies that are more vocal in their environmental claims often become the first targets of journalists and activists. Delmas and Burbano (2011) suggest an example of the seafood giant Safeway that launched the 'Ingredients for Life' advertising campaign that boasted the firm's sustainability ambition. Canadian eco-activists bought a large advertisement in The New York Times with the tagline 'Ingredients for Extinction' illustrated by the photos of dead seals and sea lice in the nets—the side effect of the firm's salmon-catching floating nets. The activist campaign was successful in making the company reduce the floating nets that generated risks for the sea life and increased pollution.

The main issue with greenwashing is obfuscation. The embellished corporate claims usually contain a grain of truth—which is not easily detected in the voluminous and decorated reports and pledges of a company, institution or a brand. Challenging the greenwashing requires meticulous attention, time and scientific literacy to test the validity of promises. Some of the emerging scholarly recommendations include teaching circular economy to the business and management university students (Kopnina, 2018)—a practice that can be equally helpful in journalism training—or trying to build an AI-powered detector for checking the greenwashing assertions en masse (Cojoianu et al., 2020). Legal initiatives and policymaking are picking up speed, yet are slowed down by the very same hurdles (Lyon & Montgomery, 2015)—the challenges of unpacking the granular cause-and-effect processes in production and distribution of goods; measuring the scale of proposed green innovations. Combatting greenwashing also requires a level of confidence and intensity that can be on par with the force of greenwashing itself. This brings media professionals to face the next big threat to the public discourse on climate change—the merchants of doubt.

Doubt-Mongering, Diluted Terminology and the Carbon Footprint Calculator Paradox

Both single brand campaigns and industry-wide greenwashing are overpowered by the grander issue—what Naomi Oreskes, a Professor of History of Science at Harvard University, calls the doubt-mongering (see Oreskes & Conway, 2010). For decades, industries such as tobacco, fossil fuels and fertilisers have engaged in seeding doubt in the public discourse. The strategies of doubt are aimed at challenging science—for instance, the studies that show that smoking is harmful, or that fossil fuels contribute greatly to the climate change (Oreskes, 2022). The techniques of doubt-mongering include cherry-picking (zooming in on one point of uncertainty and blowing it out of proportion); misrepresenting scientific evidence; funding research to deflect attention from the primary reasons of climate change (i.e. investing in the contentious research of ideologically biased think tanks); questioning the credibility of scientists; and insisting on prioritising personal responsibility of consumers over industry or specific corporations (Oreskes, 2022: 29–31).

Oreskes and Conway (2010) compare the practice of funding dubious researchers or think tanks to the construction of 'Potemkin villages'. It is a reference to the old Russian fable of the high-level bureaucrats in Tsar Russia that would create the showcase happy villages when the Empress was about to pay a visit to the area. In the 1980–90s, this strategy was tested by the tobacco companies who coined the Council for Tobacco Research (it originally featured 'Tobacco Industry' in the title but the communication professionals advised the funders to drop 'industry') (Oreskes & Conway, 2010). The Council recruited scientists that would speak up for the 'work' of the Council and defend the tobacco industry, divert attention from and question the studies of unbiased researchers. Another tactic of the tobacco lobby comprised funding 'citizen groups' of apparently concerned members of the public that spoke against climate science or suggested waiting for more evidence.

"Doubt is our product since it is the best means of competing with the 'body of fact' that exists in the mind of the general public," a tobacco company executive once mentioned (Michaels, 2005: 96). The same approach of challenging science—and manipulating the cautious language and complexity of scientific work—has been applied to climate change. Over 30 organisations have been funded by ExxonMobil, a fossil fuel company, to misrepresent science, Oreskes (2022) attests.

Supran and Oreskes made headlines in 2017 when they published a comparison of ExxonMobil's internal reports on climate data and the company's public-facing communication, such as advertorials in the likes of The New York Times over 1977–2014. The researchers revealed that, 'accounting for expressions of reasonable doubt, 83% of peer-reviewed papers and 80% of internal documents acknowledge that climate change is real and human-caused, yet only 12% of advertorials do so' (Supran & Oreskes, 2017). The predominant framing of ExxonMobil's advertorials was 'doubt' (81% of advertorials took this position), along with 11.5% displaying 'acknowledge' position and 7.5% 'acknowledge and doubt'. 'ExxonMobil contributed quietly to the science and loudly to raising doubts about it'—Supran and Oreskes (2017: no page) conclude.

Catchy terms are another persistent issue of greenwashing. While it is understandable that reducing climate change to relatable terms and actions is a logical human move, the attractive simplicity of some of the tools and common terms risks undermining the purpose of addressing climate change.

It is not by chance that one of the most common tools of measuring sustainability among non-scientists is the carbon footprint calculator. Individuals can input the amount of meat they eat, the number of flights they take, the source of energy they use—and their individual emission footprint pops up on the screen in front of them. This number apparently helps to tweak individual behaviours and become more planet-friendly—at the very least, this is impression the tool is likely to give.

British Petroleum, a fossil fuel company that changed its name to the short and unassuming 'BP', was the force behind the carbon footprint calculator (Kaufman, 2021; Solnit, 2021). Brought to life by the creatives of Ogilvy & Mather, a legendary advertising agency, the carbon footprint calculator was launched in 2014. Ever since, the carbon footprint calculations have been featured in a myriad of respected media publications, brand communication and used by the governmental agencies (Kaufman, 2021). The allure of carbon footprint calculator lies in the combination of gamification of the emissions assessment, the simple efficacy it promises (e.g. cut your flights, eat less meat and the figure on your calculator will go down), as well as appealing to the competitive spirit within humans. Comparing footprint among friends, colleagues and brands turns using it into a self-growth device as well as a social activity; it outsources consciousness and critical thinking, rewarding the users with the sense of pride or clarity in return.

While reducing individual greenhouse gas-emissions footprint is a noble and useful activity indeed (Kaufman, 2021), it should not divert the attention from the country-wide and industry-wide emissions that have to be cut in the first place. Having said that, carbon footprint calculator makes for an excellent greenwashing tool—many social corporate responsibility initiatives proudly calculate the emissions saved by the rather timid acts of changing old lightbulbs for the energy-efficient ones, introducing car sharing and solar panels on the office rooftops. The crisis is big, but small actions make people feel better about themselves and their companies—it makes the efforts quantifiable and presentable in corporate reports. Yet this tool remains completely toothless if other issues and sources of emissions are not addressed.

To prove that, the real-life experiment no one asked for brought the world to a halt. The Covid-19 pandemic in 2020–21 introduced lockdowns over the planet and inflicted a drastic reduction in flights, car rides and productivity of the factories many of which had to shut down to accommodate social isolation of the workers. Despite the dramaticism of the situation, the global emissions only took a light dip—and went back up once the business as usual resumed (Kaufman, 2021). This has shown that cutting individual emissions is a commendable act but brings no real change in isolation. Similarly, a class at the MIT University calculated that even a Buddhist monk or a penniless student are responsible for high-level emissions as long as they live in a country where basic provisions such as roads, energy, police, libraries and courts are enabled by the 'dirty' energy (Chandler, 2018). Even a homeless person that eats in soup kitchens and sleeps in homeless shelters still produces a sizeable amount of greenhouse gas emissions—albeit indirectly, as long as the fossil fuels are a base of the energy system of a country (Chandler, 2018; Kaufman, 2021).

While the term 'carbon footprint' has already entered the public and media vocabulary, there are climate scientists that insist on not amplifying the term enabled by some of the biggest emitters out there. A Harvard scientist Geoffrey Supran who researched ExxonMobil (as cited in Kaufman, 2021: no page) stresses that instead of inspecting the individual carbon footprint, people should use 'fossil fuel emissions' or 'fossil fuel footprint' as the measuring tool.

Similar to 'carbon footprint', 'carbon offsets' has been another seemingly progressive term that is aimed to drawing people's attention to the climate impact of companies and offering them tangible ways of mitigating it. What started off as a clear, understandable idea—a high-emitting business, such as an airline, for instance, could buy offsets that mean investment in reforestation or eco-transport in remote lands—became tarnished by greenwashing (Polonsky et al., 2010). Carbon offsets has become an industry of and in itself, with McKinsey, a consultancy firm, estimating its worth as 50 billion dollars by 2030 (Schendler, 2021). According to their analysis, fewer than 5% of carbon offsets projects removed carbon dioxide from the atmosphere (Ibid.).

'Offsetting is worse than doing nothing', a UK-based climate researcher Kevin Anderson (2012) writes. He argues that the scientific legitimacy of many carbon offset initiatives is lacking and the efforts in this field most likely increase the overall emissions, not curb them. Many scientists join the chorus of voices that warn against focussing on offsets—it removes the urgency to curb emissions now. It also fosters the illusion that offsets are a magic pill or a plaster that can mend the problem. However, many solutions are dynamic, such as tree protection or reforestation. If a wildfire strikes the area, the trees will be lost, observes Roberto Schaeffer, a Professor of Energy Economics at the Federal University of Rio de Janeiro (as cited in Simon, 2023). The validity and impact of many climate offsets are limited, while offering contentment and undeserved green credentials to the emitting companies.

To stress the point further, David Ho, a Professor of Oceanography at University of Hawaii at Manoa, proposes that focussing the discourse on offsets and carbon-capture technology is ungrounded optimism at best and dangerous deflection at worst. The technology to capture greenhouse gas emissions from the air is in its early stages—it is far from ready, from being affordable, scalable and implementable right now (Ho, as cited in Simon, 2023). Yet this technology already forms a significant part of the techno-optimistic climate discourse and verges on greenwashing, when referred to as a realistic solution to the climate issues.

Greenwashing is a tangle of truths, half-truths, half-lies and outward lies. The complexity and sophistication of the promises to act in favour of the planet make them incredibly hard to unravel. The more resources a brand or a company have for hiring the communication professionals, the more likely their version of events is to stick to the public imagination. Even the most obvious of the culprits—fossil fuel companies—have proven to be successful in diverting public attention from multi-layered action and reduced the issue to individual responsibility and guilt. The efforts needed to tackle greenwashing and doubt-mongering will require the ambition of policymakers, resources of the newsrooms and a big push to improve climate literacy and circular economy teaching at all levels of public education. Perhaps the biggest ambition lies in solving the puzzle—how to celebrate pro-climate actions and initiatives that matter and not give false hope to the society? How to distinguish meaningful green acts from the wishful thinking—often genuine—on behalf of companies and industries? Last but not least, this chapter follows the lead of Oreskes and Conway (2010) who argued for addressing the certainty of greenwashing claims with the certainty of scientific community. The volume has to go up on the precise and accurate academic findings—the media professionals can help scientists to identify the strong messages hidden within cautious language of scholarly writing, and amplify the truths. It is a big task for journalists, honest public relations and advertising professionals, government and industry regulators to educate their ranks and amplify the voices of reason and evidence, not false promises.

> **Box: Greenwishing**
>
> *'Use a pestle and mortar' to make your pesto sauce more sustainable, as opposed to using an electric blender'*—an article in a respectable national UK newspaper cheerfully suggested. 'This is going full medieval', one of the comments below the line philosophically mused. Another article from the same broadsheet recommended choosing the restaurants with no napkins, no paper menus, with smaller portions so there is less food waste; digital iPads were hailed as a paper-free way to show customers what's on offer.
>
> These instructions on how to go greener in your everyday life are not just naïve—they are misleading. It is different from 'greenwashing'—the deliberate practice of clouding the truth about the product's or company's environmental footprint. The phenomenon of greenwishing denotes the media storytelling angle that focuses on the minutiae of lifestyle adjustments at the expense of real impact and scale. It wishes well, but it misses the deep understanding of the context and weight of the suggested individual-level solutions. Even if all the humanity started grinding their own spices by hand or went to the restaurants with no paper napkins, the climate change and emissions would not stop. Same for choosing a restaurant based on it offering iPad menu (that still generates emissions) or textile napkins—there are more influential choices one can make, and reducing meat intake is one of them.
>
> How to know the difference between helpful, awareness-building lifestyle journalism and greenwishing? The line is blurred. Lifestyle media cover an important area in people's lives—their leisure time, relationships, emotions, consumer rights and hobbies. This area must be inclusive of climate change issues—in fact, it is a healthy approach to talk about climate in everyday, trivial materials, not just in breaking news and worrisome features. Lifestyle climate coverage can act a might efficacy-booster—even if people are still coming to grips with how to tackle climate change on various levels, taking small steps allows them to feel in control of a small field. Reducing meat intake, not buying new clothes—preferring second-hand—and using a train instead of a plane are some of those non-trivial behaviour changes that many people can make. Implementing these changes in people' lives means changing their way of thinking about climate—it becomes a real part of their everyday choices. People make small sacrifices—and it makes them more interested to see

changes on other levels—such as demand from their elected politicians to phase out fossil fuels and introduce a levy on fast fashion, reduce excessive consumption and invest in recycling and green energy, among other things.

One big distinguishing matter that separates healthy eco-lifestyle journalism and greenwishing is the comprehension of context and big picture. A story on replacing paper menus with digital tablets shows lack of research into digital waste and electricity use. Some advice to shop lentils from one brand and not from another needs to come with the disclaimer that this change makes sense only if a person adopts a meat-free and pulses-heavy diet. The naïve consumption adjustments shape the core of greenwishing—something to be filtered out by the more experienced editors and sub-editors trained in climate science. Perhaps one question to ask as a 'greenwishing' test—how big a reduction in emissions would this change bring on a planetary level if everyone adopted it?

Bibliography

Anderson, K. (2012). The inconvenient truth of carbon offsets. *Nature*, *484*(7392), 7–7.

BBC (2017, Novembe 19). London buses to be powered by coffee, *BBC*. https://www.bbc.co.uk/news/uk-england-london-42044852

Bousso, R., & Nasralla, S. (2023, March 16). Shell rules out more ambitious goal for end-user emissions. *Reuters*. https://www.reuters.com/business/environment/shell-rules-out-more-ambitious-goal-end-user-emissions-2023-03-16/

Chandler, D. (2018, April 16). Leaving our mark. *MIT News*. https://news.mit.edu/2008/footprint-tt0416

Cojoianu, T., Hoepner, A. G., Ifrim, G., & Lin, Y. (2020). Greenwatch-shing: Using AI to detect greenwashing. *AccountancyPlus-CPA Ireland*.

Delmas, M. A., & Burbano, V. C. (2011). The drivers of greenwashing. *California Management Review, 54*(1), 64–87.

Denisova, A. (2021). *Fashion media and sustainability*. University of Westminster Press.

Gallicano, T. (2011). A critical analysis of greenwashing claims. *Public Relations Journal, 5*, 1–21.

Gillespie, E. (2008). Stemming the tide of green wash: How an ostensibly greener market could pose challenges for environmentally sustainable consumerism. *Consumer Policy Review, 18*(3), 79.

Kaufman, M. (2021). The carbon footprint sham. *Mashable*. https://mashable.com/feature/carbon-footprint-pr-campaign-sham

Kopnina, H. (2018). Teaching circular economy: Overcoming the challenge of green-washing. *Handbook of engaged sustainability: Contemporary trends and future prospects*, 1–25.

Lyon, T., & Montgomery, A.W. (2015). The means and end of greenwash. *Organization and Environment*, 28(2), 223–249. https://journals.sagepub.com/doi/10.1177/1086026615575332. Accessed on 17 February 2022.

Michaels, D. (2005). Doubt is their product, Scientific American, June, 96–101.

Morrison, S. (2017, November 20). London buses to be powered by leftover coffee from today, *The Evening Standard*. https://www.standard.co.uk/news/transport/london-buses-to-be-powered-by-leftover-coffee-from-today-a3695661.html

Nemes, N., Scanlan, S. J., Smith, P., Smith, T., Aronczyk, M., Hill, S., Lewis, S., Montgomery, A. W., Tubiello, F. N., & Stabinsky, D. (2022). An integrated framework to assess greenwashing. *Sustainability*, 14(8), 4431.

Oreskes, N. (2022). The trouble with the supply-side model of science. *Proceedings of the Indian National Science Academy*, 88(4), 824–828.

Oreskes, N., & Conway, E. M. (2010). Defeating the merchants of doubt. *Nature*, 465(7299), 686–687.

Polonsky, M. J., Grau, S. L., & Garma, R. (2010). The new greenwash?: Potential marketing problems with carbon offsets. *International Journal of Business Studies: A Publication of the Faculty of Business Administration, Edith Cowan University*, 18(1), 49–54.

Romero, P. (2008, September 17). Beware of green marketing, warns Greenpeace exec, *ABS-CBN News*. https://news.abs-cbn.com/special-report/09/16/08/beware-green-marketing-warns-greenpeace-exec

Schendler, A. (2021, August 31). Worrying about your carbon footprint is exactly what big oil wants you to do. *New York Times*. https://www.btlt.org/wp-content/uploads/2021/12/Opinion-_-Worrying-About-Your-Carbon-Footprint-Is-Exactly-What-Big-Oil-Wants-You-to-Do-The-New-York-Times.pdf

Siano, A., Vollero, A., Conte, F., & Amabile, S. (2017). "More than words": Expanding the taxonomy of greenwashing after the Volkswagen scandal. *Journal of Business Research*, 71, 27–37.

Simon, J. (2023, March 5). Climate solutions do exist. These 6 experts detail what they look like. *NPR*. https://www.npr.org/2023/03/05/1160783951/6-scholars-explain-what-a-real-climate-solution-is

Solnit, R. (2021, August 23). Big oil coined 'carbon footprints' to blame us for their greed. Keep them on the hook, *The Guardian*. https://www.theguardian.com/commentisfree/2021/aug/23/big-oil-coined-carbon-footprints-to-blame-us-for-their-greed-keep-them-on-the-hook

Supran, G., & Oreskes, N. (2017). Assessing ExxonMobil's climate change communications (1977–2014). *Environmental Research Letters, 12*(8), 084019.

Tan, E. (2017, November 20). Shell and Bio-bean are powering London buses with coffee, *Campaign*. https://www.campaignlive.co.uk/article/shell-bio-bean-powering-london-buses-coffee/1450536

Thin, S., & Lewis, N. (2020, May 14). Energy drink? A UK startup is turning coffee into fuel, *CNN*. https://edition.cnn.com/2020/05/14/business/coffee-grounds-recycling-gec-spc-intl/index.html

CHAPTER 5

The 'Ignorance as a Choice' Paradox, and the Role of Depleted Resources in the Responses to Climate Messages

The dichotomy between 'knowledge' and 'ignorance' takes roots in the Enlightenment. Knowledge and ignorance have been culturally constructed notions through centuries—and now, with the ubiquitous influence of social media, clickbait media stories and overwhelming 'information obesity', knowledge on climate change becomes a part of identity construction. With fluctuating and hybrid identities of modern citizens (Bauman, 2013), climate change awareness becomes just one more element to adopt or to reject in fluid identity construction.

Martin Bauer advances the 'knowledge-ignorance' paradox as an important concept of social sciences in modern times. Despite pockets of very well-informed people in a society, the majority are assumed ignorant on scientific issues. Professor of Sociology Sheldon Ungar (2000) goes further as he proposes ignorance as the starting point of the conversation of the public knowledge of climate change.

Is Ignorance a Route to Happy Life?

Ignorance was not enjoying good reputation throughout the centuries. Lack of literacy and scarcity of curiosity have been associated with the blocks to civilisation, with mere disempowerment. Ignorance used to be synonymous with laziness. However, the tide has turned for ignorance in the second half of the twentieth century.

© The Author(s), under exclusive license to Springer Nature Switzerland AG 2025
A. Denisova, *Effective Climate Communication*,
https://doi.org/10.1007/978-3-031-67340-5_5

'Ignorance is bliss' as a saying is attributed to the eighteenth-century Thomas Grey who, in turn, was inspired by the Ancient Roman thinker Publilus Syrus. A Syrian, he came to Ancient Rome as a slave, but gained respect and liberation from his master, carving himself a career as a writer. Publilus famously coined 'In knowing nothing, life is most delightful' maxim, which is ironic given that his intellect saw him out of exploitation.

Another view of ignorance is explored by faith studies. Since the Medieval times, knowledge was seen as a path to understanding God and His governance of the world (Abdalla, 2022). Yet even the twelfth-century philosophers came to the realisation that language and science fall short of comprehending God's true reality. A Jewish philosopher of the thirteenth century Isaac Albalag declared that individuals should not be ashamed of the comprehension that they are incapable of coming close to God's divinity. If they are wise, they should be happy with ignorance.

Moreover, in Christian tradition, in the Gospel, the apostles St Paul and St Peter are presented as two variations of a good man. One is brilliantly educated and influential, while the other is reaching towards God in his simplicity and lack of scholastic training.

Fast forward to the 2020s, and some people stick to **ignorance as a rational choice**. Up to 40% of people choose wilful ignorance over full knowledge (Vu et al., 2023). There are at least four reasons why one may wilfully subscribe to the 'ignorance camp'.

1. It can be done out of self-protection—identifying as ignorant in certain spheres helps to shield an individual from accountability. This is especially prominent in the areas where knowledge implies responsibility.
2. It can be evoked as a liberating tool to enjoy the social and physical environment without thinking about potential dangers (Bauer & Joffe, 1996; Vu et al., 2023).
3. It can be a self-defence against worrisome, anxiety-triggering information. Recent research in the UK demonstrates that people get increasingly disengaged with the news as they see it as a cultivator of anxiety and doom (Lorenzoni et al., 2007; Woods et al., 2018). Not knowing becomes a self-preservation, or even a self-care, choice.
4. Ignorance can be invoked as a diagnosis to the society that limits the agency of certain social groups and largely excludes them from decision-making. Women and youth are more likely than men to talk about the feeling of distance from the public issues (Joffe & Farr,

1996). They experience the inability to impact the public realm and thus tend to equate 'knowing' with being able to affect anything. It becomes a social identity.

A popular scientific view in the early 2000s suggested 'the deficit of knowledge' as the main obstacle to people's awareness of and eagerness to act to mitigate climate change (Bauer et al., 2007). This 'deficit model' was later debunked as it became clear that other factors are to blame for the lack of action: these include people's ideologies, selective attention and psychological resistance (see Nabi et al., 2018, for an overview). Rational appeals have proven to be limited in their capacity to inspire and motivate people. Furthermore, a correlation has been found in the preference for wilful ignorance and reduced altruistic behaviour (Vu et al., 2023).

'Information deficit model' has also been prominent in policymaking (Whitmarsh, 2009). The idea that people just lack information to make rational decisions has been floating in media effect studies for a while. It has been disproven by many further studies (see Irwin & Wynne, 1996; Whitmarsh & O'Neill, 2011) that demonstrated that people are not a heterogeneous entity, and various individuals and groups have various values and resources, thus making one-size-fits-all knowledge distribution approach very limited. Knowledge alone does not drive change.

Pivot to Emotions—Why a Tonic of Fear and Hope Is Desired for Climate Action

This is when the role of emotions came to the forefront of environmental, psychological and communication studies. The pivot to feelings allowed to acknowledge that the emotional lens through which people experience climate change messages—worry, anxiety, hope, frustration, interest—largely affects their support or neglect of climate policies. The emotional reaction was responsible for a 50% variation in rational decision-making about endorsing climate action. It has proven to be more potent than sociodemographic categories or background in influencing people's perception of climate change (Smith & Leiserowitz, 2014, as cited in Nabi et al., 2018: 444).

The affective responses of fear (Hart & Feldman, 2016) and hope (Ojala, 2012) are often identified as the potent forces of engaging the audiences with climate awareness and action. Fear, as any strong potion, comes with side effects when used in large doses—it may motivate some people to act, but will cause the reaction of resistance, source denigration or problem minimisation in others, as a coping mechanism (Nabi et al., 2018). Hope may seem purer as a feeling—it can trigger the enthusiasm to participate in climate action, especially if a constructive range of actions is suggested (Ojala, 2012). Nevertheless, as ethereal and optimistic as hope may seem, it may still be bearing dark undertones—if the hope arises from the dismissal of the magnitude of the crisis, then the hope-induced action may brink on complacency.

To mitigate the shortcomings of two strong sentiments and advance the debate, Nabi (2015) proposes the emotional flow concept. This model connects the range of emotions and estimates the response. A message of fear that contains hope **and efficacy components** is more likely to motivate individuals to support climate action (Nabi et al., 2018).

Comparable to watching a Netflix blockbuster, engaging storytelling often takes us on a journey. Thinking of any underdog story—whether Slumdog Millionaire or The Queen's Gambit—one would notice the classic storytelling structure. Dating back to Aristotle and mythology, a three-act arc suggests the manipulation of emotions in the audience. A hero is called to action—motivated by a threat or a need to help those weaker than her—she rises to the occasion, but fails again, overwhelmed by the magnitude of the task and her own weaknesses—yet she manages to persevere and achieve a greater success after all. When the audience follows this classic narrative, they are taken on an emotional rollercoaster. It is challenging and frustrating on the surface—fear motivates the initial steps, yet it is the hope and the emerging sense of self-efficacy that keeps the character going and also maintains our involvement with the story. The interplay between fear, hope and empowerment is perfectly controlled by the storyteller and is rather satisfyingly recognised by the audience.

Same framework can be applied to the climate messaging. If one could orchestrate the emotional response that leads to the third act—the culmination, the emotional release and the satisfaction from the participation in a fulfilling act of storytelling—then climate communication can do the same. Nabi's (2015) emotional flow model arouses the reader by the message of fear, but then provides them with the avenues of hope by offering efficacy solutions. How to achieve the right balance of a message being worrisome enough but allowing space for solutions? How to avoid the illusory hope and point to the real issues lying underneath as well as solutions available?

The components of a climate story vary greatly—some point to the damage already done, and others warn of the issues yet to come. Would people react better to the message that helps them to maintain the status quo and minimise losses from the climate change? Or would they be interested to know what they can gain, if engaged in climate action? Apart from the actual gains to one's business—say, increasing the harvest from crops that are adaptable to extreme weathers—these gains might include reputational benefits, satisfaction from helping the community or engaging in altruistic activity (Titmuss, 1970).

The research into the framing of loss-gain has not brought conclusive results. Even though the influential figures of Daniel Kahneman and Amos Tversky found in the 1980s that the frame of loss was more motivating for the people to take action, later research indicated otherwise—much depends on the topic, context, efforts and affordances. The limits of human rationality become obvious when applied to the decisions related to risk. Kahneman and Tversky's (1984) proposition that averting a risk motivates people to act has been challenged by many scientists in the following years. Neuroscientist Tali Sharot (2011) found that people act slower when faced with a risk than with a gain. She brings forward one experiment as an example. Participants were given either of the two tasks: to press a button to win a dollar each time, or press a button to avoid losing a dollar. Those expecting to win were faster to act than those faced with the potential decrease to their fortune. Sharot stresses that people need a clear reward to motivate action—and in this case, gained dollars are more exciting than lost money. Applied to climate change, people need to see the benefits from engaging with the pro-climate behaviour—whether to feel the sense of belonging to a community of like-minded people, or to do the socially desired thing, or to act in alignment with their values. The benefits must be clear and attainable to make the motivation strong enough.

Confirming the classic instinctive response to threat—'fight, or flight, or freeze'—people in the dollar experiment were more likely to stay passive in the face of a danger. Their response was closest to 'freeze' rather than 'fight' or 'flight'. To mitigate this, Kahneman (2011) suggests the approach of routine to make people less hesitant in decision-making. If something becomes familiar and less imposing, people can act quicker and at a lower cost to their psychological well-being. Kahneman makes an example of a financial trader—these people make lightning-speed decisions worth millions every day of their working life—and they lose and win significant sums of money while they do. In this environment, gambling large sums of money becomes 'business as usual', or what Kahneman defines as 'broad framing'. For climate action, presenting pro-climate behaviours as common sense and social norm makes it easier for more people to make an environmentally friendly choice at lower decision-making costs.

A counterpoint to 'broad framing' perspective is that ignoring climate change and finding excuses not to act can also become 'business as usual'. This attitude can be facilitated by the increased emotional discomfort or doubt triggered by the climate news—this means that media professionals should urgently measure the coverage that provokes despair, not action. Not every empowerment is speedily possible—and surely the journey to efficacy (Bandura, 1977) is rarely linear. Sharot (2011) points to the complex nature of stress that prevents action. She argues that, after traumatic events such as disasters, wars, terrorism acts, or market collapse—experienced first-hand or lived through news consumption—people can feel pessimistic and withdrawn.

When it comes to climate action, the exposure to the messages about gains and losses can create mixed reactions in the audience. There is no clear consensus whether the threat of the potential loss is more motivating than the promise of a potential gain.

In some cases, the possibility of, say, losing the Amazon Forest generates a visceral reaction of clicking, sharing and engaging with the message about it (Carbon Brief, 2023). In this case, the weight of the potential calamity is so big, and the physical and symbolic properties of the Amazon are so imperative that the reaction is strong. However, clicks and shares of the story about the Amazon Forest at peril do not mean engagement with climate action in the long run.

A different example relates to the perceptions of the climate change effects on public health. A study of semi-structured interviews was held with the adults in the US. They perceived as negative the messages about the harmful consequences of the climate crisis on health. They did, however, find the climate change issue more relatable through the lens of health. This reaction of fear was contrasting to the reaction to other storytelling—on the policies that mitigate the risks and lead to health benefits. The coverage of institutional action was perceived positively and optimistically. What is helpful in this study of fear and hope, action and inaction, is the mixed outcomes. It was revealing that some behaviour change solutions—i.e. reducing the meat intake—were faced with resistance and counterarguments. However, other solutions—introducing a better public transport infrastructure, clean energy, non-polluting cars—were met with interest and approval. This example proves the complexity of balancing out wins and sacrifices, rational decision-making and opting for the 'ignorance identity'. Framing climate change not as a problem, but as the domain of 'solutions and the many co-benefits' (Maibach et al., 2010) is therefore suggested as the way forward.

'The Ignorance Explosion' Paradox and Heuristic Bias in the Context of Accelerated Societies

Since Immanuel Kant, it is deemed prestigious to be knowledgeable because this reputation brings power. Knowledge removes stagnation and advances progress (Bauer & Joffe, 1996). Even in mental health therapy, the knowledge of oneself is presented as a route to a more balanced life and self-realisation.

Bauer and Joffe (1996) see a paradox in how distribution of knowledge affects the 'statistics' of the whole population. They claim that the deeper educated people go into their respective fields and the more they discover, the harder it is for the general population to catch up. This creates a paradox—there is plenty of complex knowledge in areas, but that makes the average knowledge levels lower, compared to the glowing extremes.

This does not entail praise for the scientists as beacons of balanced knowledge. Bauer, Durant and Evans (1994) take this argument as far as suggesting that even those specialising on specific areas have big fields of ignorance otherwise. One can be a brilliant astronomer and a poor budget planner. An excellent economist might be lacking in their understanding of climate science.

It was observed a while ago—well before the boom of the Internet and social media—that the intellectual capacity of humans is very limited. The increasing speed of life in the modern, technology-heavy 'accelerated societies' (Rosa, 2013) leaves the humankind overwhelmed when facing challenges of pollution, destruction of natural resources and societal conflicts. Despite plenty of science and knowledge available to man to make an impact in these areas, humans are struggling to understand the core of them and to concentrate on resolving them (Lukasiewicz, 1974).

Professor of Mechanical and Aerospace Engineering, Julius Lukasiewicz (1974) ironically called this phenomenon of attention shortage and knowledge evaporation 'the ignorance explosion'. The memory of a man is known to be able to increase by only 7–8 times. This storage space is partly occupied by short-term and operational memory—the ones crucial for making decisions and solving problems.

Furthermore, human cognition is constrained by another limitation—'channel capacity', identified as the ability to distinguish and pass on the information received (Lukasiewicz, 1974). A modern human is attacked by an increasing number of stimuli—from work and family requirements to the pinging updates on social media, notifications from professional media and apps, endless news and feature articles on everything in the world. In this oversaturated environment, one retains their biological capacity at a rather modest level. Approximately seven stimuli at once are identified as the maximum that a person can identify and transmit (Lukasiewicz, 1974). This can be called 'a stock of knowledge' (Rosa & Scheuerman, 2009). It defines the amount of knowledge that is useful in the present moment.

The Nobel laureate in Economics Professor Daniel Kahneman (2011) agrees with this focus on limited reasoning capacity. He explains that our thinking happens on two main levels—the fast and the slow. He calls them System 1, 'automatic system', and System 2, 'effortful system' (Kahneman, 2011: 29–30). The first one relies on quick reactions and decisions, often based on stereotypes, bias and emotions. The second one involves self-control, attention and spending more mental capacity on comprehension and analysis.

The psychologist Roy Baumeister has demonstrated that effortful thinking is tiresome and depletes the overall mental energy. This explains why, when people are tired or overwhelmed by their daily tasks, their 'system' has less capacity to think straight, analyse critically and make rational judgement. As Kahneman (2011: 41) puts it, drawing on various psychological experiments, 'you would be more likely to select the tempting chocolate cake when your mind is overloaded with digits. System 1 has more influence on behaviour when System 2 is busy and has a sweet tooth'. It is important to acknowledge in the light of climate change engagement that people with busy and tired minds are more likely to make selfish choices or make biased judgements.

Yet another limitation to knowledge storage and capacity for critical thinking lies in a different area entirely—ideological affiliations. They strengthen the shortcut thinking, the urge to jump to conclusions relying on ideological predisposition as a scaffolding for information processing. American sociologist Dan Kahan calls the average capacity to understand numbers and graphs 'ordinary science intelligence' (as cited in Hayhoe, 2021: 52–53). He measured to what extent people were able to understand scientific facts, statistics and analysis. The results were interesting—those with higher scientific literacy were a little more likely to agree that climate change is the result of human action. Those with lower science intelligence were two times less likely to agree with the same point. Yet the picture changed dramatically when Kahan looked at the ideological affiliations of the participants—those with democratic views were positive that climate change owes to human activities; the conservatives were mostly sceptical about the human causes of the crisis.

German sociologist Hartmut Rosa (2013) adds another challenge to the already complicated discussion on the many layers of judgements, rational and irrational thinking. He views not just individual human beings as stretched beyond capacity, he evaluates societies as nauseatingly accelerated—due to technological, communication and social changes. A Professor of Sociology, he states that we are experiencing a paradoxical shrinkage of time—the narrow area between the past and the future is a 'slipping slope'. We have barely processed the past and we are already halfway into the future (Rushkoff, 2014), being swiftly carried forward as if on a fast tide in a slope in a water park. The abundance of information and ideas—delivered at a speed of a Japanese bullet train—also means there are sources of information and misinformation out there that allow anyone to confirm their bias, as misjudged as they may be (Sharot, 2011).

However, among the solutions to a man's limited memory and thinking capacity is the coding system. We all rely on language, laws, iconography to guide us through rules, communication, education and societal life. Lukasiewicz (1974) argues that coding helps to outsource some of the knowledge, yet humans should be able to have enough capacity to remember what the symbols mean and how to decode them.

In climate change discourse, the shortcuts are emerging—'1.5 degrees', 'net zero', 'carbon footprint', etc. They are helpful beacons of the big topic and yet need to be universally understood and decoded to make climate change a household conversation. These codes should not become a myth in themselves—further deliberation on the benefits and issues of symbolic discourses like that of the 1.5 degrees is offered in Chapter 7.

What does emerge as a clear trend in this complicated analysis is the limited resource of humans to deal with knowledge. Understanding of climate crisis needs time and consideration. It can rely on symbols as shortcuts—but those have to be easily understood and universally decodable. Ideological bias and depleted systems have a further constraining effect on the engagement with climate knowledge of the tired, overwhelmed or agency-deprived citizens.

Eureka, or How Human Psyche Chooses the Quick Answer

Daniel Kahneman (2011) and his long-time collaborator Amos Tversky discovered the mental shortcut that people tend to exploit when faced with a big question. Instead of trying to grasp the immense complexity and analyse all possible ways to answer it (which is impossible), our mind substitutes the difficult question with a more doable one. For example, a complex question 'How much would you contribute to save an endangered species?' is replaced with 'How much emotion do I feel when I think of dying dolphins?' (Kahneman, 2011: 97–106).

This simpler question is called 'heuristic reasoning' (Kahneman, 2011: 98), and the term derives from the ancient Greek 'eureka', which roughly translates as 'I have found it!'. This mental shotgun allows the mind to jump to a simpler substitute and produce an answer.

In the case of climate change, the target question may be 'Do you care about climate change?'. The heuristic question that pops up in your mind may be 'Do I like the coral reefs enough to make sacrifices?' 'Or am I afraid enough to give up meat?' 'How much do I care in order to take action?'.

What affects the choice of the heuristic question? Sometimes it is the context or even the order in which questions are asked. Kahneman (2011) makes an example of a university experiment when students were asked whether their happiness correlates with their dating life. Surprisingly, in one experiment there was no correlation at all, while in the other one, with exactly the same questions and similar respondents, the happiness was closely tied to dating. What was the difference? The order of questions. In the first scenario, the students were asked:

How happy are you these days?

How many dates did you have last month?

The students showed almost zero correlation between good mood and dating. In the second scenario, they were asked about the number of dates first and about happiness second—and they were much more emotional and reported a perceived link between the two. Similar experiments that asked questions about relationship with parents and then happiness released similar results. One's emotional priming was playing a big role in the way they perceived the big question. The choice of the heuristic question was massively significant and changed the course of thinking about the big question.

WHY PEOPLE DENY CLIMATE CHANGE, OR THE 'DRAGONS OF INACTION'

Driving a car to buy burgers. Flying to holiday destinations using low-cost airlines. Working for the industry that relies on fossils fuels or consumerism and overproduction. These are just some of the comfort-driven reasons why people would madly defend the 'status quo'. The reasons for climate denial are multiple and overlapping—some cognitive, some psychological, some a mixture or an ideological mantle.

This section identifies the three main reasons why people deny, delay or diminish climate change. The theories juxtaposed lie at the crossroads of social psychology, sociology, media studies and political studies.

1. **Uncertainty**. Canadian psychologist Robert Gifford came up with the term 'seven dragons of inaction' to explain the prominent barriers to pro-climate action and behaviour change. They differ from the structural barriers—say, lack of public transport infrastructure in the area where people are constricted to use cars—and point to the more subtle, layered psychological reasonings. They comprise limited cognition; ideologies; comparison with the others; sunk costs; mistrust; perceived risks; and limited behaviour (Gifford, 2011).

   While the ignorance paradox has been discussed before—knowledge is helpful but is never enough to motivate for sustained action—other reasons come into play when it comes to cognition. Emotional overload can be accompanied by uncertainty—the term nods to the ambiguity people feel over the kinds of action or impact they may have. They may end up taking this as a—subconscious—excuse for inaction (Gifford, 2011; see also de Kwaadsteniet et al., 2007). The 'social dilemma' theory is helpful here—it covers the conflict between personal interests and the interests of the group. This 'conflict' may be over resources or behaviours. There are various factors involved that make people willing to pursue selfish interests or cooperate—and uncertainty has been discussed as an impeding factor for group action and equality (de Kwaadsteniet et al., 2007).

Media reporting can contribute to this uncertainty—Gifford (2011) points to the surprising issue with meticulous scientific language in climate studies and stakeholder reports. Where scientists wrote 'likely' and 'very likely' as the measures of probability for certain temperature spikes and disasters to happen, the general public interpreted those estimates as vague and worthy of dismissal (Budescu et al., 2009, as cited in Gifford, 2011: 292). The perceived uncertainty led to the audience underestimating the magnitude of the issue and finding a shortcut excuse to not take them seriously.

2. **System justification.** Ideological bias may include political affiliation, but they can also simply arise from the comfort of the existing lifestyle. 'System justification' (Feygina et al., 2010) refers to the commitment to protect the status quo and dismiss the urgency of actions needed to mitigate climate crisis. Comfort justification grows stronger when scaffolded by political conservatism and nationalism. The tendency to look for stability is inherent in human beings—we don't like change and we don't like to be forced to change. People seek security and reassurance, as well as social ties and acceptance of those living in the same community or social system (Jost & Hunyady, 2005, as cited in Feygina et al., 2010: 327). The positive side of system justification means the reduction of anxiety, uncertainty and fear in those who practice it; yet the ugly side is that those who are in the privileged position are more likely to ardently defend the existing system despite its shortcomings (Feygina et al., 2010). Curiously, sometimes even those harmed by the system may be willing to engage in justification when they think that this position is most likely to lead them to stability and safety, when any alternative is seen as a worse-off scenario.

As a remedy, Feygina et al. (2010) suggest integrating climate action into the narratives of patriotism and status quo ('protecting our land'). This is where media communication can shape the vision of stability and prosperity as united with pro-climate behaviours; the eco-friendly behaviours can therefore be presented as socially and ideologically desirable, rightful acts for the country and fellow countrymen.

3. **Social comparison.** Since the seminal studies of Leon Festinger (1954) who proclaimed that people are social creatures who depend on comparing themselves with those they perceive to be equal or superior members of the society, the social comparison remains a strong reason for action or inaction. Comparison with the others helps to distinguish social norms—i.e. predominant attitudes, beliefs and behaviours (Cialdini & Jacobson, 2021). Closely linked to the social proof theory (see more in Chapter 6), the social comparison principle underlines how people mine social norms from the behaviours of the others (Gifford, 2011). The effects of people's looking around and modelling their own behaviours on the perceived 'social norm' have been seen in the examples of tax evasion, sun cream use, bullying and fruit consumption, among others (Cialdini & Jacobson, 2021).

There is a paradoxical relation between the 'globalised' and 'provincial' norm perception. While social media and traditional media often emphasise the developments in wider societal norms, the more localised the application of the norm is, the easier it is for the person to follow suit. In various experiments, it was clear that people are more motivated to convert to the climate-friendly behaviour when it had been explained to them how others—similar to them, living in close proximity—have applied that behaviour (Ryoo et al., 2017, as cited in Cialdini & Jacobson, 2021: no page). This means that, when people are primed to see other members of their community or the aspirational members behave in a certain eco-friendly way, they would be more likely to adopt this behaviour. As stressed throughout this book, assigning blame and telling people that climate neglect is the prominent social norm is a really poor and unhelpful practice—this would drive further complacency rather than any change. Humans are social animals and need role models and social proof as adjustment prompts.

The issues of knowledge deficit and ignorance as a choice are closely intertwined with the formation of identity in a fluid, changing world. The media hold the power to advise people in this environment and speak to their values and affiliations without sacrificing climate change action. More attention is encouraged to the views and behaviours of conservative members of the public—and those whose livelihoods depend on the real or imaginary reliance on the world built on fossil fuels.

Bibliography

Abdalla, B. (2022). *The bliss of ignorance: A twisted view.* University of Birmingham, news. 4 April. https://www.birmingham.ac.uk/news/2022/the-bliss-of-ignorance

Bandura, A. (1977). Self-efficacy: Toward a unifying theory of behavioral change. *Psychological Review, 84*(2), 191.

Bauer, M. W., Allum, N., & Miller, S. (2007). What can we learn from 25 years of PUS survey research? Liberating and expanding the agenda. *Public Understanding of Science, 16*(1), 79–95.

Bauer, M., & Joffe, H. (1996). Meanings of self-attributed ignorance: An introduction to the Symposium. *Social Science Information, 35*(1), 5–13.

Bauer, M., Durant, J., & Evans, G. (1994). European public perceptions of science. *International Journal of Public Opinion Research, 6*(2), 163–186.

Bauman, Z. (2013). *Identity: Conversations with Benedetto Vecchi.* John Wiley & Sons.

Carbon Brief. (2023). *Analysis: The climate papers most featured in the media in 2022.* Multiple authors, 5 January. https://www.carbonbrief.org/analysis-the-climate-papers-most-featured-in-the-media-in-2022/#:~:text=Across%20the%20top%2025%20papers,Science%20with%20three%20papers%20each

Cialdini, R. B., & Jacobson, R. P. (2021). Influences of social norms on climate change-related behaviors. *Current Opinion in Behavioral Sciences, 42*, 1–8.

de Kwaadsteniet, E. W., van Dijk, E., Wit, A., De Cremer, D., & de Rooij, M. (2007). Justifying decisions in social dilemmas: Justification pressures and tacit coordination under environmental uncertainty. *Personality and Social Psychology Bulletin, 33*(12), 1648–1660. https://doi.org/10.1177/0146167207307490

Festinger, L. (1954). A theory of social comparison processes. *Human Relations, 7*, 117–140.

Feygina, I., Jost, J. T., & Goldsmith, R. E. (2010). System justification, the denial of global warming, and the possibility of "system-sanctioned change." *Personality and Social Psychology Bulletin, 36*(3), 326–338.

Gifford, R. (2011). The dragons of inaction: Psychological barriers that limit climate change mitigation and adaptation. *American Psychologist, 66*(4), 290.

Hart, P. S., & Feldman, L. (2016). The impact of climate change–related imagery and text on public opinion and behavior change. *Science Communication, 38*(4), 415–441.

Hayhoe, K. (2021). *Saving us: A climate scientist's case for hope and healing in a divided world*. Simon and Schuster.

Irwin, A., & Wynne, B. (Eds.). (1996). *Misunderstanding science? The public reconstruction of science and technology*. Cambridge University Press.

Joffe, H., & Farr, R. (1996). Self-proclaimed ignorance about public affairs. *Social Science Information, 35*(1), 69–92.

Kahneman, D. (2011). *Thinking, fast and slow*. Macmillan.

Kahneman, D., & Tversky, A. (1984). Choices, values, and frames. *American Psychologist, 39*(4), 341.

Lorenzoni, I., Nicholson-Cole, S., & Whitmarsh, L. (2007). Barriers perceived to engaging with climate change among the UK public and their policy implications. *Global Environmental Change, 17*(3–4), 445–459.

Lukasiewicz, J. (1974). The ignorance explosion. *Leonardo, 7*(2), 159–163.

Maibach, E. W., Nisbet, M., Baldwin, P., Akerlof, K., & Diao, G. (2010). Reframing climate change as a public health issue: An exploratory study of public reactions. *BMC Public Health, 10*(1), 1–11.

Nabi, R. L. (2015). Emotional flow in persuasive health messages. *Health Communication, 30*(2), 114–124.

Nabi, R. L., Gustafson, A., & Jensen, R. (2018). Framing climate change: Exploring the role of emotion in generating advocacy behavior. *Science Communication, 40*(4), 442–468. https://doi.org/10.1177/1075547018776019

Ojala, M. (2012). Hope and climate change: The importance of hope for environmental engagement among young people. *Environmental Education Research, 18*(5), 625–642.

Rosa, H. (2013). *Social acceleration: A new theory of modernity*. Columbia University Press.

Rosa, H., & Scheuerman, W. E. (Eds.). (2009). *High-speed society: Social acceleration, power and modernity*. Penn State University Press.

Rushkoff, D. (2014). *Present shock: When everything happens now*. Penguin.

Sharot, T. (2011). The optimism bias. *Current Biology, 21*(23), 941–945.

Titmuss, R. (1970). *The gift relationship: From human blood to social policy*. Vintage Books.

Ungar, S. (2000). Knowledge, ignorance and the popular culture: Climate change versus the ozone hole. *Public Understanding of Science, 9*(3), 297.

Vu, L., Soraperra, I., Leib, M., van der Weele, J., & Shalvi, S. (2023). Ignorance by choice: A meta-analytic review of the underlying motives of willful ignorance and its consequences. *Psychological Bulletin, 149*(9–10), 611.

Whitmarsh, L. (2009). Behavioural responses to climate change: Asymmetry of intentions and impacts. *Journal of Environmental Psychology, 29*(1), 13–23.

Whitmarsh, L., & O'Neill, S. (2011). Introduction opportunities for and barriers to engaging individuals with climate change. In L. Whitmarsh, I. Lorenzoni, & S. O'Neill (Eds.), *Engaging the public with climate change: Behaviour change and communication* (pp. 1–13, 2012). Routledge.

Woods, R., Coen, S., & Fernández, A. (2018). Moral (dis) engagement with anthropogenic climate change in online comments on newspaper articles. *Journal of Community & Applied Social Psychology, 28*(4), 244–257.

CHAPTER 6

From Emotions to Determination: The Communication Tools for Free Riders and 'Conditional Cooperators'

People are slow adopters of new ideas and behaviours. Psychology scholars argue that we need a 'social proof' of the novel lifestyle or social action before we fully get on board. This social proof entails recognising that people from our circles adopt a new behaviour.

For instance, second-hand shopping used to be a niche choice of ethically minded consumers, left-wing ideology proponents, supporters of charity shops, as well as a preference of some parts of the British upper-class society. However, with the uneven financial capacities, rising climate awareness and the market looking for new ways to sell goods, new digital platforms emerged—the likes of Depop, Vinted, Ubup, eBay offer modern ways of doing just that—acquiring fashionable or necessary items, worn or unworn, at a lower price, from fellow consumers. Across Europe, second-hand shopping is gaining momentum as the manifestation of 'modest consumption' (Steffen, 2017), 'anti-consumption' (Lee & Ahn, 2016), the 'not poor, but shop clever' approach (Gregson & Crewe, 2003) and the solace of those struggling to afford mainstream consumption becoming 'excluded customers' (Williams & Windebank, 2002) who revel in the affordability of second-hand shopping.

This change of mindset—that second-hand is cheap but also chic—did not happen in one day. Scholars see the adoption of novelty through a myriad of theoretical paradigms: from the Theory of Planned Behaviour, Theory of Interpersonal behaviour, Motivational Model, to Diffusion of

© The Author(s), under exclusive license to Springer Nature Switzerland AG 2025
A. Denisova, *Effective Climate Communication*,
https://doi.org/10.1007/978-3-031-67340-5_6

113

Innovation and Social Cognitive Theory. What emerges as a pattern in the analysis of these concepts is that values alone cannot shape behaviours but are a potent factor to consider in combination with other triggers and variables.

Do attitudes shape behaviours? In the late 1960s, the jury was out—and then Allan Wicker from University of Wisconsin reviewed (1969) the existing research and proved that no, attitudes do not transform into behaviours, the correlation was incredibly weak. The response to this finding was the range of theories stemming from sociology, psychology and digital media studies.

The Theory of Planned Behaviour, originating in the 1980s, takes into account three main components that indicate the inclination to engage in a certain behaviour (Ajzen, 1991). These factors are attitude, subjective norm and control over the behaviour. An example of that can be the interest in second-hand shopping among teenage girls and young adults in the UK. A girl may see the discussions about second-hand items in her peer group's social media channels and distinguish this behaviour as a social norm. One-third of all 16-to 24-year-old UK residents are present on the fashion reselling platform Depop (Levy, 2021). Does it mean all of them have given up fast fashion and unethical brands for good?

Well, no. Theory of Planned Behaviour is limited in its prediction powers. So many other factors can affect the behavioural choice of our proverbial teenage girl—it may be the high or low dependence on group approval; economic and psychological constraints on adopting a specific behaviour (the lack of time, money or effort to browse multiple offerings on Depop; the uneasy relationship with used clothing—finding it unhygienic, challenging to buy without trying things on or worrying about potential scams); lack of trust in group solidarity—assuming that peers still shop in fast-fashion outlets all the same. Low behaviour intake may also be due to negative experiences that people might have had on second-hand platforms before.

Interestingly, some studies suggest that the social norm, attitudes and intention are strong predictors of the self-interest behaviours, but not the altruistic ones. In this case, Theory of Planned Behaviour is more potent when it comes to the direct benefit for the person. In a study of Chinese rice farmers (Zhang et al., 2020), it was established that their attitudes and social norm were useful predictors for the uptake of adaptation behaviours. They were rather willing to explore a more resistant crop or change the patterns and calendars of planting—these adaptation techniques were beneficial for their business. They believed that other farmers would adapt similar approaches.

Proving the theory right, the attitudes and social norm have proven less influential when it came to altruistic behaviours of these Chinese rice-growers. When asked about the intention to help mitigate the effects of climate impact—such as trying the reduction in crop intensity or forsaking the fertiliser—they were less inclined to act. The main issues were the perceived lack of control over mitigation techniques—and the weaker influence of personal values. Those farmers with pronounced altruistic and pro-environmental beliefs were more likely to engage in mitigation techniques. This study (Zhang et al., 2020) highlights the limitations of the Theory of Planned Behaviour yet opens the debate on values and norms—and how the media and storytelling around climate change may affect values, acknowledgement of the emerging social norm and, in turn, behaviour changes.

Not just the social norm, but personal norm can be influential in the adoption of new behaviours. The Value-Belief-Norm (VBN) theory looks at the role of personal norms and values on active adoption of behaviour. An influential study that applies the Value-Belief-Norm to environmental support was done by the Washington-based Paul Stern and his colleagues in 1999. It postulates that the inclination to support an eco-movement is borne out of three overlapping factors—personal norms, values and beliefs. Personal norms may include those of family loyalty, respect for the elders, serving the community—in order to tap into these, social movements or media messages need to create the sense of obligation (Stern et al., 1999; see also Snow & Benford, 1992). If a media message succeeds in appealing to these, it is more likely to motivate people for action.

The term 'values' in the VBN theory, when applied to environmental concerns, often implies altruism. This vision stems from Shalom H. Schwartz's model of basic values (1973, 1977)—the American social psychologist and his colleagues proposed ten basic values: Conformity, Tradition, Security, Power, Achievement, Hedonism, Stimulation, Self-Direction, Universalism and Benevolence. From the pro-climate perspective, the most important values driving action might be universalism and benevolence.

The core value of universalism suggests that many people are acutely aware of the consequences of their actions and societal developments by people outside of their immediate bubble. This awareness and empathy generate the sense of responsibility. Schwartz (1973, 1977) famously coined the 'ascription of responsibility' concept that explains how one feels responsibility for the events affecting others or does not assume that responsibility. Therefore, appealing to personal values is paramount in campaigning as much as in media communication. Snow et al. (1986) refer to this approach as 'value amplification'. The most successful campaigns match the political values of a candidate to the personal values of the prospective voters. In environmental communication, strong media narratives on climate action rely on the accentuation of the values and beliefs (perhaps, conservation, or Tradition, and prevention of threats, or Security) that can appeal to the core values of the audience.

What if one does not wish to assume personal responsibility for the issue as mammoth as the climate change? Free riders can be convinced—through the pressure of beliefs. When it comes to beliefs, the public awareness of the scale and impact of climate change becomes crucial. A person needs to believe that the material and immaterial aspects or habitat important to personal values are under threat—this can trigger action. There can be a dozen of values that create a potent framework to motivate action—they can vary from universalism to fundamental religious beliefs, from hedonism and freedom to social justice and achievement. The Values-Norms-Beliefs theory is helpful for the study of behaviour adjustments and climate change because it assesses the responses of each of the three components to the stimuli, and how they affect practical action. It has been studied, for instance, that the norms play a strong part in consumer behaviour and willingness to make sacrifices, but do not affect one's readiness to join a demonstration. Values, however, can fill this gap—universalism, tradition and openness to change have been identified as stouter motivators for becoming a part of a collective mobilisation (Stern at el., 1999).

Bringing the Hearts to the Table, Not Just the Minds

The question of values is important—and these can be communicated on various levels and with varying levels of scientific intensity. A renown climate scientist and communicator Professor Katharine Hayhoe urges 'to bring our hearts to the tables, not just our heads' (2021: 33). She is adamant that people are not rational and bombarding them with facts, figures and statistics will not convert minds to care about climate change.

Neuroscientist Tali Sharot (2011) agrees with that—evolutionary development did not make our brains appreciative of facts for decision-making, we are still driven by the instincts and beliefs that are important to us. This means people will take out the proverbial 'swords' to protect their values, beliefs and even bias.

In fact, the smarter and more educated an individual is (and may even be great at maths!), the more rational arguments they would find to protect their bias. Their intellect and defence system will be capable of producing a string of immaculately logical arguments that would justify their beliefs and choices. This psychological bias is called 'motivated reasoning' (Hayhoe, 2021). A person chooses the facts and arguments selectively, in order to match their values and beliefs, and seeking to confirm what they already know. Emotions come first, and reasons follow.

These studies confirm the uncomfortable truth—we are driven by emotions first and foremost, and the rational mind, or 'slow thinking', as Kahneman puts it, tends to take the backseat. How to communicate climate effectively? Knowledge has a role to play in climate discourse, but must be rooted in shared values, motives and identities (Hayhoe, 2021; Sharot, 2011).

A South African journalist Lameez Omarjee, part of the Oxford Climate Network, agrees. Omarjee (2023) quotes an interview with a climate scientist who posited: 'Find something that people care about and explain to them why climate change is going to put that at risk'. These cultural values may differ between countries and communities—for South Africans, for instance, it is the heritage sites within national parks that bear the memories of the ancient civilisations—these sites are at the high risk of floods and climate-related destruction.

Hayhoe (2021) makes a strong case for finding shared identity grounds, passions and items people care about to be able to start a conversation about climate change. She argues that Christians can connect based on the Bible that presents the Earth as the dominion of humankind, which comes with the responsibility to curate and treat well what God has given to us. For parents, climate action means preventing many crises that might affect the generations to come. For children, climate action may mean the preventive measures to address the risks to the health of elderly parents, multiplied by the climate crisis. Whether one is a diver, a surfer, an artisan, a fashion designer, a foodie or a wildlife aficionado, climate coverage has to appeal to the interests and values of the audience—and only then climate topic can be broached.

The Gift of Climate Action

How can we overcome biology, the selfish genes that make us care about ourselves and our kin, as well as our habits, tastes, our luxuries before the planet? Is there a secret mechanism within every person that can be activated for the social good? Yes and no. The survival instincts have taught us to compete over resources and distrust others. But the very same instincts make us social species, attached and willing to do well for our 'tribe'.

The classic studies on altruism explain how the care for the distant others at the expense of own time and resources it is not a natural, primal reaction to the life stimuli, but the product of our consciousness and a fruit of civilisation. Altruism entails caring attitudes and behaviours that expand beyond our immediate kin and reciprocal relationships.

Richard Titmuss, the father of altruism studies, famously wrote about the motivations of blood donors in the UK and the US in the book *The Gift Relationship: From Human Blood to Social Policy* in 1970. He distinctively declared that altruism is not a biological reaction but a social construct. When identifying the motivations that drive people to give blood to strangers, he found that many were compelled by a health issue they or someone close to them have experienced before, and how they felt the urge to 'give back'. A hugely important **other** motivation for altruistic behaviour was the 'warm-glow feeling'—it is the pleasing emotion that a giver experiences as a reward for a selfless deed (see Andreoni, 1990, 2001). This means that contributing to the social good comes with internal benefits such as the pride of being a good citizen, the reaffirmation of one's identity as a good person and potentially religious or spiritual bliss.

Can altruism be pure, liberated from any expectation of the thrilling satisfaction and moral rewards? Pure altruism is a debated—for instance, Kahneman and Knetsch (1992) state that there is always some sort of satisfaction behind one's charitable activity. Being a social construct, altruism and philanthropic behaviour in all its forms can tell a lot about the society. By studying the socially praised and accepted charitable practices, and attitudes towards charitable behaviour, one can draw a picture of the social capital in this particular society.

'Social capital' (Coleman, 1988; Putnam, 1993) refers to the social cohesion that prosocial behaviour helps to accomplish. Societies with a healthy social capital are better equipped in overcoming inequalities, bringing social groups together and encouraging citizens to care not only about themselves, but the community. Titmuss (1970) found that altruistic prosocial behaviour greatly assists social cohesion. In fact, Titmuss was the first one to affirm that relying on gift relationship of voluntary donations (UK scenario) is safer and more economically viable that living the issue at the mercy of monetary rewards and management by for-profit organisations (US style).

Social capital is related to the question of social norms. Altruistic tendencies are not enough to ensure stable and consistent climate action within the population—unless altruism becomes *the* social norm, which is unlikely. Yet the inquiry into the warm-glow niche of altruistically minded behaviour allows to identify six predictors of pro-climate action. These are patience, willingness to take risks, altruism, trust, positive reciprocity and negative reciprocity (Andre et al., 2021). Some of these predictors of climate action speak to the altruistic tendencies within humans and their belief in the collective responsibility. Moral universalism—the idea that we are accountable for the people and living conditions beyond our immediate society—is another strong influence on climate action. Yet again, it is far from being a prevailing thought or norm within any country or population. Other predictors of climate action include responses to the wrongdoings—for instance, the willingness to punish pollution, waste of resources and similar climate offences.

In summary, personal values and beliefs need to be strengthened by the perceived social norm and social capital in a particular society. The media outlets have the power to construct the social norm and encourage a stronger sense of mutual responsibility and social cohesion that are closely related to the virtues of altruistic giving and pro-climate behaviours.

In for a Ride? The Moderation Techniques for Free Riders and 'Conditional Cooperators'

How do people know about the norms of the others? How do we learn the 'average' in the public opinion? It was Georg Wilhelm Friedrich Hegel, a nineteenth century's German philosopher, who once called reading a newspaper an alternative to a morning prayer. Throughout the twentieth century, the media acted as one of the main institutions to bring people 'on the same page' in terms of agenda setting, gatekeeping and information provision. Now, with the abundance of sources of information, analysis and opinion, a new tool is quietly emerging as a potent reminder of who we are as a society, what we believe in and how we act. A humble tool with formidable capabilities.

Public Opinion Surveys

One of the major obstacles to the growing sense of social efficacy is the distrust in the others. People are not willing to sacrifice their international travels, steaks and shopping if they are not convinced that their neighbours—close by and faraway—will do the same.

People tend to underestimate the climate intentions and actions of the others. The media can turn the tide on this. The research on 8,000 US adults (a representative sample) in 2021 (Andre et al., 2021) demonstrated that most people vastly misconstrue the climate actions and beliefs of fellow countrymen. The authors argue that much more information on the majority opinion should be dispersed to the population to increase the public support of costly initiatives—such as carbon tax—and assure people that they are more alike than they think.

Homo sapiens are 'conditional cooperators'. Swiss Professor of Behavioural Economics Urs Fischbacher and his colleagues (2001) have determined that about 50% of people are 'conditional cooperators', meaning that their decisions to contribute to public good were dependent on the actions of the others. These peer-dependent actors wanted to see the evidence of others contributing first. Furthermore, this experiment—albeit small, of only 44 students—also revealed that around 30% of participants remain free riders. These participants were not participating in contributing to the public goods no matter how other people acted.

The position of 'conditional cooperator' may be an independent solidified psychological stance, or a variation of a behaviour that can show signs of altruism, reciprocity and fairness elsewhere. Fischbacher et al. (2001: 11) concluded that even unconditional cooperators may demonstrate a tendency of 'self-serving direction' at times; the do-good tendencies weaken over time. This does not mean that altruistically minded people or those dependent on conformity deteriorate into egoism—but rather that the willingness to contribute to the common good is not a status quo; it is unstable and evolves in an uneven trajectory.

Despite the small sample and bold conclusions, the findings of Fischbacher and his colleagues have stood the test of time. A review of about 17 studies that replicated the original experiment (Thöni & Volk,

2018) revealed the same pattern. Most people are conditional cooperators (over 60%), every fifth person is a free rider (20%), and only under 5% are unconditional do-gooders.

'Pluralistic ignorance' is a helpful term that explains the social norm in transition—it indicates the perception of one's own norms as deviation, minority stance, while assuming that the majority holds a different social norm. This concept elucidates the discrepancies between private beliefs and the evaluation of the others. This confusion is exacerbated by the fact that most members of the society could be holding similar personal beliefs but behave in the vein of what they perceive as a 'social norm' even if it is contrary to the truth. In 1924, Floyd Allport, one of the founders of social psychology studies in America, first identified the illusion of universality—he noted that members of the group assume that the attitudes of the others are uniform. This notion has since been theorised and checked by a myriad of studies.

Pluralistic ignorance can create dangerous social tendencies—Latane and Darley (1970) bring forward the example of bystanders. They explain how, if a victim of an emergency is lying on the floor in the middle of the street, the passers-by have the choice whether to help or assume that nothing bad happened. In this case, the people are unsure whether the situation is serious or not, but looking at other people pass by and mind their business, they may conclude that there is no need to help the affected person. The erroneous judgement on the beliefs of the others results in inaction. People internalise what they believe to be a social norm.

Another experiment strikes closer to home for many academics. In the famous research on classroom dynamics, Miller and McFarland (1987, 1991, as cited in Prentice & Miller, 1993: 2) established a scenario in which a professor delivered a complex lecture. He then asked students to pose questions. The classroom went silent. Why was that so? Many students were afraid of asking a question and appearing stupid. Simultaneously, they assumed that—if others did not raise their hands—their peers had probably understood the material. Pluralistic ignorance in this experiment left students feeling inadequate and, worse even, alienated and different from their classmates.

Another experiment—at Princeton University—checked the stereotypical assumption that students are keen on alcohol consumption and wild parties (Prentice & Miller, 1993). Despite the issues that this may cause for their health, relationship and studies, students reported a strong belief that others consume plenty of alcohol and enjoy the drinking culture. It was established as the norm. Nonetheless, it was illuminating to see the personal views on alcohol—many were not comfortable with the drinking convention but estimated the view of the majority as highly comfortable with it. The figure was somewhere in the middle in the estimate for friends—respondents rated them halfway between their personal beliefs (drinking is not great) and the assumed social norm (drinking is awesome). Men were more likely to adjust their behaviours to the perceived social norm, while women stuck to their principles more often. The discrepancy between one's idea of the social norm and private views was fuelling alienation (Prentice & Miller, 1993).

Any educator reading this may recognise the pattern—when your personal tutees tell you that they are not making as many friends and not enjoying as many parties 'at uni' as they were hoping to, they widely assume that their peers are having the times of their life. The difference between the assumed norm and private ideas can therefore be painful and confusing—even if the difference is illusionary.

When applying pluralistic ignorance to climate change awareness and willingness to act, the media have a massive role to play—they may create the prevailing norm of climate awareness and climate action. Or it can disrupt the good efforts by suggesting doom and gloom and pushing people towards inaction. How do people identify the social norm? Traditional media, social media, popular culture, as well as advertisements, hospitality venues and shop windows have a role to play.

In the current media environment, increasing the understanding of the prevailing social norm as pro-climate is crucial. There are colossal numbers of people who care about climate change—statistics puts climate worry high cross many global surveys (three-quarters of population in the UK—ONS, 2022; 79% in France—Gross, 2018). Increasing the cognisance of the social norm can be done through the dissemination of public opinion results, offering more public space and airtime, and popular storytelling, to pro-climate communicators; encourage social media owners to critically examine their practices and make sure that climate denial is not featured in the newsfeeds, while pro-climate influencers are enjoying fair advantage in algorithmic selection. Moreover, it is paramount to keep endorsing pro-climate beliefs and actions as default in the TV series, fashion campaigns and consumer communication, and in similar tenets of broader culture.

Box: Public opinion polls—the low-key intervention mechanism
The effect of low-key information intervention—showing to the people the results of the public opinion shared by their fellow citizens—is incredibly potent. Authoritarian countries have learned the trick a while ago—they produce bogus surveys of the public opinion that praise the status quo and thus suppress any dissent. Individuals remain under the impression that the majority is supporting the government, and any disagreeing voices are a weak minority. This prevents people from seeking to establish communities, networks and see action. Kuran (1991, as cited in Andre et al., 2021: 8) proposes that the misconception about the beliefs of the others slows down the collapse of restrictive regimes.

In democratic countries, however, with independent and legitimate sociological surveys, conducting research on a representative sample of the population is a fruitful technique. Communicating the results to a wide populace is a clever way to boost the morale and induce climate action. Andre et al. (2021) specifically suggest informing the climate deniers that they are a minority—as it has proven to increase their intention to support climate policies.

If you're reading this from the UK, pause for a moment to estimate how many of your fellow citizens worry about climate change and consider it the major concern. You will be surprised.

According to the ONS (the reliable Office for National Statistics, 2022), 74% adults reported being worried about climate change, and 74% nominated it as the biggest concern. Only the rising cost of living ranked higher on the list of the concerns (at 79%).

Three in four adults in the UK made some lifestyle changes to tackle climate change. The unforeseen and reassuring component of the national survey was that—across all levels of education and class—the climate worry remained strong. It was higher among women and those with a degree (around 80%) and slightly lower among those with college, or school qualifications or less (66–71%).

If you're reading this from the US, fear not—62% of the Americans try to tackle climate change, and around 80% believe action should be taken to address it (Andre et al., 2021).

How do we make sense of the myriad of events happening around us? One of the founders of social psychology Floyd H. Allport proposed the concept of structuring, organising ideas in a coherent narrative in the public eye. Widely distributing the results of the opinion polls assists in solidification of the impression of universality—the feeling among group members that their peers think and act in the same way (Allport, 1924).

Similarly, the renown Canadian American sociologist Goffman calls the connection of ideas with the sense of self 'ordering'. Goffman gives an example of everyday actions and behaviours as a direct reflection on the sense of self. Both 'ordering' and 'structuring' point to the need of some choreography of social interactions—an act of one person leads to an expected reaction from another person, and this continuation of events and domino effects is performed anew every day. Allport, in particular, found in one of his experiments that, when workers were not allowed to communicate during the day, their productivity increased and their ideological views skewed towards the centre, which Allport concluded as the search for the common sense or status quo. It did not prevent the scholar from rejecting social structure and focussing on the 'events' instead, arguing that people remain individuals even when they connect in collective agency (Czarniawska, 2006).

In summary, ideas do not travel in the vacuum; they borrow from the previous ideas and previous events. People create expectations on social interactions from the ideas they encounter in popular narratives; they shape and reshape their beliefs and behaviours according to the ideas and interactions they engage with. To take this further, in my other research (Denisova, 2019) I devoted half a decade to the studies of memes, the cultural artefacts similar to genes, which convey information and form subjectivities, delivering ideas from one sender to another, from area to area, from context to context. Ephemeral storytelling is inseparable from meaning-making storytelling. Viral fragments of communication, conventional interactions and public opinion surveys structure the reality—this allows people to perceive the order of events as the 'system' of comprehension of the world, and act accordingly with this view.

SOCIAL PROOF AS THE FOUNDATION OF SOCIAL NORM CONSTRUCTION

Late 1980s. Imagine you are leaving a nightclub and stumble upon a bin—full of identical cans of an energy drink. Then, on another night, you see it again—all bins around the trendiest nightclubs in town are full of empty energy drink cans. Same brand. All the cool crowds are enjoying it, why not I?

This was the famous campaign by the Red Bull—instead of persuading the club goers to try their fizzy drinks to improve their energy levels and make them dance the night away, they applied the opposite strategy. They have made people believe that everyone else was 'in the know', and they were not.

Humans are social animals, we like to conform and don't like to be outliers. People like to be socially accepted and have an inherent need for belonging (Baumeister & Leary, 1995). The concept of 'social proof' is essential in social psychology as it delves into this unique human trait. Robert Cialdini, the Professor of Psychology and Marketing (Arizona State University, Stanford University), came up with this theory in 1984, observing that people evaluate the social codes, behaviours and conventions to determine the appropriate behaviour. 'Multiple others and similar others' determine the way we think, look and behave—Cialdini postulated (2007). This means that a message is more likely to influence people if it implies that the majority is doing or thinking this already.

While Red Bull famously simulated popularity to become essentially popular in the context of recreational culture and nightlife habits (Nguyen, 2021), many academic studies further prove the 'social proof' concept.

Cialdini's (2001) experiment involved asking the guests of the hotel chains to reuse their towels. Reusing these linens in guest bathrooms can save plenty of water and electricity, yet it has been hard to convince visitors to sacrifice this additional comfort for the sake of the environment. Instead of signs 'Do it for the environment' and 'Do it for your children's sake', Cialdini proposed the signs that claimed that most guests reuse their towels. This approach, based on the 'social proof', has proven to be more efficient in encouraging people to reuse the towels.

Cialdini (2001: 100) defines social proof as the approach that impacts people's perceptions of certain actions as 'correct in a given situation to the degree that we see others performing it'. It has been confirmed as an efficient strategy in gaining compliance (Cialdini, 2001; Cialdini et al., 1999), convincing people to donate to the charity (Reingen, 1982), returning a lost wallet in a big city (Hornstein et al., 1968), making erroneous mathematical estimates when in a group (Asch, 1951), among other experiments.

Social proof is so powerful that it can distort people's perspective. Solomon Asch (1951) famously found in his experiments that people were unable to measure the length of a line on a card correctly when the majority responded differently. They would trust the majority more than their own eyes. Festinger's (1954) social comparison theory provides another angle to this—it postulates that people have a strong tendency to compare themselves to those similar to them. Festinger's (1954) and Asch's (1951) studies apply meaningfully to the influence of social media on beauty standards, lifestyles and perceived values among social groups. It can also stimulate public awareness and action intention for the benefit of climate change mitigation.

Even small prompts can generate effective results. Simply hearing the laughter of the others at a cinema, at a comedy club or when watching a sitcom with the pre-recorded audience laughter makes us enjoy the humour more and find the jokes funnier than if watching on our own (Fuller & Sheehy-Skeffington, 1974). The social laughter theory assists the understanding of social cohesion and applies to many mundane social situations—for instance, the convivial politeness and small talk with a smile are common in public places in many cultures of Europe, Japan, US, to ensure a liveable and unthreatening social coexistence.

The more similar is the demographics that serves as the basis of the 'social proof' narrative, the higher the compliance of the others in a similar social stratum. In a classic experiment with lost wallets in New York, people were more likely to return the wallet if they have heard of someone like them—another born-and-bred New Yorker—return the wallet. The compliance was lower when the example of a person returning the wallet used a foreigner—at least this what the mood was like in the 1980s when the study took place (Hornstein et al., 1968).

Social proof can manifest with slight variation in different cultures. An experiment comparing student compliance in the US and Poland revealed the Poles to be more impacted by the social proof, the reminder of the actions of the others. This was understood as the effect of the collectivist tendencies within the society, as opposed to the more individualistic Americans (Cialdini et al., 1999). Yet e-commerce platforms suggest that whether individualistic or collectivist, users are likely to be nudged by the evidence of others behaving in a certain way. This is why the clever marketing techniques of the hotel booking websites as well as fashion platforms utilise the 'social proof' techniques to lure the customers towards a purchase. They may casually inform a site visitor that '15 users have this in their basket' or demonstrate some assurance that the item is 'trending in the last 24 hours', even though these claims may not be true.

People favour gradual changes—or, academically put, enjoy the allure of the dynamic norm navigation. The dynamic norm is the changing behaviour norm of other people over time (Sparkman & Walton, 2017). This can entail a gradual adaptation of the established social norm to the new times, small steps towards the development of a new norm. An example of the dynamic norm communication can be telling the coffeeshop customers that 'more people use their own coffee cup rather than single use cups'. Gradually, over time, this statement helps to make a reusable cup a new norm (a 17% increase of the use of practical cup—see Loschelder et al., 2019).

The expectation that the new norm will reach the pantheon of the established social norms is called 'preconformity' (Sparkman & Walton, 2017). It refers to the rising assurance among people that more and more people will behave in a newly established way in the future—e.g. eating meat once a week, using train instead of flying and buying second-hand. One caveat to preconfirmation is that the individuals who are already pursuing a new behaviour are under the impression that more people are behaving—or will soon be behaving—like them (Ibid.). This may generate a social gap between the groups that are keenly following the new social norm, and those that are not aware of it or have not got the means to adopt it. It reduces the space for a dialogue about the emerging norm.

Media play a massive role in communicating the dynamic norm—they help to increase the expectation that the emerging behaviour is becoming the conformity choice. 'Reusable cups versus plastic cups' is an example of the dynamic norm solidifying—both policy action and media coverage in many countries point to avoiding plastic as the new norm (Loschelder et al., 2019). The dynamic norm acceptance thrives in specific situations. A café that shows a sign that 'the increasing number of customers prefer a reusable cup' is creating a situational nudge towards sustainable behaviour. Yet it is not granted that other sustainable choices will follow in other areas of a person's activity (Sparkman & Walton, 2017; Walton & Wilson, 2018). The dynamic norm needs monetary and cultural trellis to let it organically adhere to the environments and people's behaviours.

Communication techniques are part of the cultural trellis that ensure that the new norms can be adopted and adapted gradually to the needs of the consumers and the environment. Gregory Walton from University of Stanford and Timothy Wilson from University of Virginia are proposing 'wise interventions' and 'story editing' approaches to cure social misunderstandings.

They bring forward an example of the poorly performing pupils from an ethnic minority detected in a particular school—these students respond to feedback negatively, while the teacher genuinely wants them to improve. An intervention technique is suggested to explain to the students that the feedback is coming from the position of care and aims to help pupils 'reach a higher standard'; this is supported with the reassurance that the teacher believes that a student is capable of achieving the goals of the tasks. Deemed 'high standards and assurance approach', this discourse framing has helped to bring a dramatic change to the student willingness to review their essays. Among African American pupils, the figure rose from 27 to 64%—the added note sustained the students' trust in their teachers for the rest of the school year (Yeager et al., 2014, as cited in Walton & Wilson, 2018: 618).

'Wise interventions' go beyond negative stereotypes to help find identities and storytelling plots that bring about social good, not conflict and blaming. This approach calls for a considerate implementation of mixed methods of persuasion to a specific situation. One of the health interventions, for instance, can be based on the IMB approach—information, motivation and behaviour (Fisher et al., 2014, as cited in in Walton & Wilson, 2018). This method was applied to the education on protected sex (Fisher et al., 2014; as cited in Walton & Wilson, 2018). Participants received information, which motivated them to avoid the risk; they further engaged in learning the practical ways of holding a conversation with a partner on implementing changes. The IMB technique has helped to increase protected sex as the norm around the world over time (Ibid).

In climate storytelling, the articles about the harms brought about by plastics in our households can act as a strong nudge for change along the IMB skills route. The coverage that specifies risks of consuming microplastics from cutlery, containers, bottles, as well as children's toys, provides alarming information about the issue. This coverage activates the motivation to defend the health of the reader and their family. The inflicted behaviour change may include exchanging plastic food containers for glass ones; not heating plastic in the microwave to avoid particles becoming toxic; and reducing the reliance on plastic in the kitchen and nursery. The combination of information and practical steps results in a new norm of plastic avoidance and higher awareness of the toxicity of some man-made elements in the ecosystem.

Wise interventions (Walton & Wilson, 2018) start small, testing the grounds for a successful adaptation of the nudges and behavioural changes in a modest setting, to then expand it progressively to broader contexts. This recommendation to tread lightly is helpful for media coverage too—it is important not to overwhelm readers with the recommendations on what to do, and the information on the risks they face. Too much information can cause withdrawal, too much pressure on one's values and habits at once can lead to resistance; the context and framings can change within groups and for individuals; polarisation is a real threat to collective action (Judge et al., 2023; see also Loscheldeder et al., 2019). All this means that 'wise interventions' in media coverage on climate should be reflective of the audience's response; they are advised to avoid blaming individuals, or portraying undecided or inactive citizens as villains. This approach may include the focussed IMB interventions on specific issues—e.g. plastic harm in the households—and then progress to similar topics—e.g. plastic harm in workplaces, hospitality, industries, etc. Societies are a like a ball of yarn—some loosely attached, some densely tousled threads of stories. Detangling some of the stories and weaving them in a single tread is a long task that requires meticulous attention and utmost care.

People tend to prefer convention and stability over change—hence why those who disrupt the routine face initial backlash. People resist to social change for four reasons: when they feel that the system is under threat, when they depend on the system, when they feel in little control, or when they find the system unescapable (Kay & Friesen, 2011). They also see the benefits of the existing status quo overpowering the potential benefits of the system change (Kay & Friesen, 2011). The social change advocates need to reach a critical mass in numbers to transform the social convention (Centola et al., 2018). From this viewpoint—deriving largely from game theory—the group needs to have reasonably high numbers but, even more importantly, high commitment levels. Is there a numerical figure for the critical mass, for the **tipping point**? The estimates range between 10 and 40% of the population, depending on the cause (Centola et al., 2018). Acknowledging the rising pro-climate awareness, climate worries and pro-climate behaviours among the population is instrumental. The more people would see themselves in the media—their class, background, ethnicity, aspirations mirrored by climate advocates on the screen and through publicly acknowledged statistics—the more inclined people would feel to continue in their eco-efforts.

Where does environmentalism sit on the Maslow pyramid of needs in the Western societies that are less affected by the immediate damage of climate change? Although a critical 1995 study of the sociologist Inglehart suggested that those with higher socioeconomic status have more time and energy to think and act on climate, the more recent research has disputed this (Dunlap & York, 2008; Mildenberger & Leiserowitz, 2017). These studies have proven that, even with economic crises and worsening of the individual or national financial conditions, people maintain their belief in the need to address climate change, disregard of class or financial capacity. The only potent trigger for abandoning or amending one's environmental beliefs is the shift in people's political allegiance.

Another ally of social psychology in tackling climate apathy is the concept of **consistency**. People like to be seen as honest, steadfast and true to their values—in other words, consistent (Cialdini, 2007). This means that when a person embarks on a journey of pro-climate action— say, buying second-hand instead of new clothes—they are more likely to explore other eco-friendly behaviours like flying less or giving up on an odd burger. This view aligns with the IMB paradigm and with Hayhoe's (2021) insight that people acquire stronger values through doing things that are pro-climate—e.g. installing a solar panel makes one more likely to envision themselves as caring about climate impact—with more action ensuing. Action first, idea second. This approach of leveraging small victories is known as the 'foot-in-the-door technique'—you ask people for a small act, like signing a petition, first. Then you come again to ask for a larger contribution or commitment.

An experiment was conducted in Israel in the 1980s (WP Carey News, 2007) to prove the consistency principle. It involved getting residents of a block to sign a petition in favour of building a recreation centre for people with disabilities. Two weeks later, the researchers came back and asked the residents to donate money towards the construction of that centre. Among those who had signed the petition earlier, 92% chipped in some money, while just 50% of the people who affronted the topic for the first time donated to the cause.

Consistency theory and social proof are well supported by Legitimisation of Paltry Donation research (Cialdini & Schroeder, 1976; Shearman & Yoo, 2007; Weyant, 1984). Also known as the 'Even a penny will help!' paradigm, it recommends asking for small steps in charity communication and thus increases the probability of a donation. In the case of pro-climate behaviours, it means that every little tweak facilitates the gradual transformation of the social norm.

For climate change behaviour change, this means encouraging people to take small steps towards the cause—with or without any clear link with the climate agenda at first. It may mean triggering an action first and propagating the idea second. The consistency approach may also mean tapping into people's existing values and actions—and highlighting the elements of pro-climate behaviour that align with those. If a person values efficiency and low cost, it might be reasonable to talk about the reselling websites for clothes and goods and discuss train prices over fuel costs for a car. If a person prioritises quality and tradition, it is worth reminding them of the nutritional benefits of the locally grown fruit and vegetables, unrivalled seasonality and the taste that comes with it. If a person worries about closing jobs because of the green transition, it is within reach to highlight the growing market for green jobs and especially the opportunities emerging locally.

Social Proof Contextualisation and Challenges Within Social Media

Both consistency and social proof are tools and characteristics of social psychology; they do not have agency. Both can be used unethically and bring social harm, too. To give a recent example, Naeem (2021) studied the reasons for the panic buying of certain goods—toilet rolls, pasta, rice, masks, sanitisers—during the Covid-19 pandemic in 2020–2021. He discovered the effect of social media on generating a social proof (possibly influenced by the algorithms of connections, trends, sensationalism) that triggered people to panic and rush to the shops. Social media in this case—empowered by isolation and disconnection from social encounters—created the common view of Covid-19 as a threat to basic supplies. While Naeem (2021) correctly emphasises that these responses may vary depending on race and class—e.g. not everyone can afford buying their way out of panic—they nonetheless prove the salience of social proof even when manifested on social media alone.

Social media are not neutral—their algorithms are hidden from the publics and most researchers. They rely on propagating the material that is likely to generate more clicks and sharing, thus keeping the users on the platform for longer and selling their attention to the advertisers

(Denisova, 2023). Various networks have various approaches to virality—yet it has been established that the market logic prevails over gatekeeping or responsibility logic, with stories and reactions trending, even if they bring no value to the social good. Awe, anger and anxiety are the main emotional triggers to make the story go viral (Berger & Milkman, 2012). Users and meaning makers, including the governments and the media, are forced to adapt to the platforms, instead of the platforms seeking to prioritise necessary and balanced information to the users (Denisova, 2021b, 2023).

> **Box: Ozone hole and the power of storytelling**
> Before the 2000s, very few media outlets were providing consistent, accurate and engaging information on climate change—the New York Times being an exception (Ungar, 2000). In this context, the ozone hole story was an exceptional event that managed to grasp public attention and public imagination for several reasons. Discovered in 1979, the depletion of ozone layer above the Antarctic was attributed to human activity in 1985. The main reason that was destroying the ozone density was the wide use of chlorofluorocarbons, particular gases found in fridges, air conditioning and aerosols. The elimination of these gases from production became therefore a goal of policymakers, campaigners and scientists—which was reflected in media coverage. By 1987, all countries in the world ratified the phase out of these specific gases in Montreal, in an unprecedented act of unity known as the Montreal Protocols.
>
> What distinguishes the ozone hole media discourse from the broader global warming agenda is the contained crisis frame. As Ungar (2000) explains, it was constructed in the media along the similar lines as disease outbreaks, like Ebola or similar—it created a moment in time when urgent danger and urgent action were very vividly depicted in the media, thus resulting in the narrative of moral panic. The concept of a 'hot crisis' (Ungar, 2000) is relevant here as it indicates a threat with an immediate effect on the members of population. Similarly, effective environmentalism (Boon-Falleur et al., 2022), mentioned earlier, applies to the media construction of the ozone hole worry—it was represented as a clear cause-effect scenario, and it was apparent how to eliminate the root of the problem with minimal sacrifice. The clarity of the campaigning goals and gains mobilised people across the world to boycott companies relying on the toxic chemicals (Rowlands, 1995).

> The timing of the ozone debacle was important too. Proclaimed as a big issue just at the end of the Cold War, it came at an opportune time for policymaking and coincided with the increased interest in climate change and activism (Ungar, 2000). To give an idea of the scale of coverage, the ozone hole issue received as much media coverage in the likes of the New York Times in the 1980s as 'climate change' as a whole, thus ensuring the saturation and continuous awareness-building of the public.
>
> Fast forward to the 2020s, the ozone layer is healing—it is on track to close the 'hole' once caused by the toxic human-advanced chemicals. It is a story of optimism for the climate change policymaking. There are media framing reasons that led to the successful signing of the Montreal Protocols in 1987. Firstly, the ozone hole crisis was explained through the media in understandable metaphors—a 'hole' or a 'crater' is much sharper than the 'global heating' or 'earth fever' allegories (Ungar, 2000). Second, it was focussed on one small set of solutions, as opposed to the numerous action points needed across the globe to address climate change. Third, the sacrifices needed to fix the ozone hole were 'trivial'—aerosols and fridges could remain in public use and discharged at the end of their short life span, to be replaced with the new non-problematic generation of technology as the next purchase (Ungar, 2000). Fourth, the visual symbolism of a 'shield' being under threat resonated with the fans of sci-fi movies and Star Wars in particular (Ungar, 2000), which therefore created an appealing and feasible path for collective efficacy and 'heroism'.
>
> Conclusively, the ozone hole issue was a perfect case of a contained crisis that could be represented through vivid, cultural resonant metaphors. It sparked citizen mobilisations, persistent media coverage, generated an array of relatable codes and symbolism, and was fortunate to come at a perfect timing when policy actors were available to look into it.

THE STRUCTURING OF HUMAN EXPERIENCE—THE DICHOTOMY OF INDIVIDUAL AND COLLECTIVE SENSEMAKING

Humans make sense of the universe through stories. From the ancient days until the present times, we are inclined to organise our living experiences and those of the others through stories. This is what Allport and Goffman identified as the organisation or structuring of events. Floyd H. Allport famously proclaimed that all social psychology is rooted in the studies of the individual behaviour, not collective choices. Even in the face of criticism from fellow scientists, he insisted that individuals react to stimuli—and stubbornly rejected 'collective spirit' and 'collective behaviour' (see Czarniawska, 2006 for the beautifully eloquent analysis of those debates). Institutions are 'phases or segments' of human behaviour, Allport insisted, allowing some acknowledgement of the influence of group environments on individuals yet remaining faithful to the idea of groups as collectives of individuals, in the first place. 'Interstructruing' (1974, as cited in Czarniawska, 2006: 1664) was the term Allport reluctantly arrived at— defining it as the force that shapes the behaviours of individuals through interactions.

Events vanish if they are not linked to other events by the individual (Czarniawska, 2006, interpreting Allport's later work). Katz and Kahn (1966) in their seminal studies that established systems theory argued that events have to reach a certain salience to form a 'structure'. A cycle of events is therefore needed to shape the expectations and a steady pattern of behaviour that informs the individuals. This is where Allport's insistence on individualistic nature of behaviour starts to crumble—as individualistic as humans are in their motivations, the concept of 'interlocked behaviours' comes into play.

Emanating from both Allport's (1962) work and that of his contemporaries (Katz & Kahn, 1966; Weick, 1979), it points to the fact that individuals shape their behaviours based on the stimuli and desires—but they must see and acknowledge the behaviours of other individuals to sustain these efforts. These behaviours of the others may be similar or complementing to the behaviour of an individual—but they contribute to building those invisible 'structures' that define the behaviours of the society members after all.

In case of climate change action, individuals need to be aware of the shared norm of climate change as a fact of life, which requires certain efforts and sacrifices from all. Individuals rely on the alignment of their individual beliefs with those of the others around them—following the systems theory, media communication creates a shared understanding of climate issues and climate behaviours. The nature of climate change is also ironically rhymed with the nature of interlocked behaviours—it is a chain reaction, a domino effect of global consumption and indulgence that generated climate crisis in the first place. Collective efforts underpinned by individual and group beliefs are of utmost importance to address it.

Humans have the urge to stabilise collective structures. Weick (1979) brings a curious adjustment to Allport's and Katz and Kahn's work. He proposes that people do not need to share aspirations and intentions as long as they perform within an established collective structure—it is the here and now that matters, the time and space that they share. Weick also propositioned the vision of collective structures as a grammar—a term borrowing from linguistics. The societies are 'energy-dependent' (Katz & Kahn, 1966)—they need individuals to keep behaving in a way that supports common identities and memberships. This resonates with Bruno Latour's famous claim that 'there is no society outside of us' (2005, p. 143; see also Czarniawska, 2006). Latour's postulate refers to the idea that we are but a collection of individuals who can be conditioned into certain communities and identities—it is through the actions, Latour notes, that we can brought into associations. This vision is cushioned by the idea of sustained action that can strengthen the ties and bridge differences.

Is it a chicken and egg scenario, do individual need to come together to create collective action, or does action induce individuals to take part in the emerging association?

How can a collective structure emerge? Karl Weick, a Michigan-based Professor of Psychology who introduced such concepts as 'mindfulness' and 'sensemaking' into common vocabulary, has the answer for this. Back in the 1970–1980s, he largely agreed with Allport and Katz and Kahn. Weick added to their debate by stressing the tendency for 'people (to) generate the environment through their actions' and through their attempts to make sense of these actions.

'Once a collective structure forms, people take steps to ensure it is preserved' (Weick, 1979, as cited in Czarniawska, 2006: 1666). Weick's sensemaking theory proposes three steps—perceptions, interpretations and actions that form the way we see the world and act upon our knowledge. This is more of a process than a linear algorithm (Weber & Glynn, 2006).

The complex nature of sensemaking is further highlighted by the *retrospectivity* of its application. According to Weick, sensemaking is akin to what Daniel Kahneman and Tversky called fast thinking and slow thinking, it is the immediate reactions that people have, followed by the more contemplative interpretations of why people acted in a certain way. '*How do I know what I think before I see what I say?*' is the classic question central to Weick's theory (meticulously distinguished by Weber &

Glynn, 2006). The behaviours and identity associations influence how we behave—but **there is a wide range of affiliations that we may be drawing upon when making decisions and acting upon them**.

Weick received plenty of citations and acclaim to his work in the 1980-90s, but a fair share of criticism too—the opponents wondered whether individual actions exist outside of the established social structures. This is why Weick's theory of sensemaking received backlash primarily for its neglect of institutions, environment and reducing human experience to 'context-free networking' (Taylor & van Every, 2000: 275, as cited in Weber & Glynn, 2006: 1639). Institutions play a significant role in structuring the roles, scripts and frames—for instance, an employer and an employee; loyalty and initiative; annual review and promotion mechanism.

Sensemaking is therefore not individualistic but inherently social. As Weber and Glynn (2006) succinctly put it, 'Who am I?' is closely related to Goffman's (1974) 'What is going on here?'. Individuals adopt identities and behaviours based on the social relations, frames they have been exposed to and become a part of.

Even the cult radical democratic thinkers Ernesto Laclau and Chantal Mouffe (1985)—the last ones to agree with 'the mindfulness' author Weick, and neoliberal paradigm in general—were broadly in agreement with the sensemaking paradigm. Laclau and Mouffe famously proposed that conflicts, tensions and agreements exist as a 'the multiformity of the social' (1985, 95, 96), but not as a society, or 'social order'. Does it mean we should not try to address society as whole to address climate change? **Yes and no—individual actions lead to organisational thinking, bringing people to associations**, some sort of ordering and structuring—of mind and of behaviours. Allport (1974: 19, as cited in Czarniawska, 2006: 1664) put it better:

> (S)ocial entity had no outside! Even if we should greatly coarsen our grain by observing it from the viewpoint of an astronaut, we still could not identify it … from an outside view. Societal phenomena are too implicit to be "encountered," too closely bound up with the meanings and actions of individuals. They consist of the interactions of such behaviours and the patternings that appear among them.

As I discuss in other chapters, pro-climate identity is a big ask. Most people stay on the side lines, partially invested, or invested occasionally in the active and consistent climate action. This does not mean they don't play a part in coming together as a pro-climate critical mass, a pro-climate majority. Humans are drawn to the stabilising structures and making sense of the events around them. This creates grounds for associations and the actions that arise from building them. Recycling, eating less meat or divesting from fossil fuel companies—**these actions, big and small, act as *identity gearshifts*** that bring people one step closer to building a sustainable pro-climate mentality. These actions do not convert individuals into a unified group—and the social psychology classics insist on their focus on individuality. Nonetheless, the acts and messages construct the reality; they assist in the 'structuring' of interactions, actions-reactions, conventions and rituals of common sense that affect the collective mind after all.

Institutions, such as education, workplace, public bodies—and the media, of course—help individuals to make sense, create meanings and form affiliations. People make sense *with* institutions, not *in spite* of them (Weber & Glynn, 2006). Community, neighbourhood and social circles are also forceful factors in shaping associations. Until an individual internalises pro-climate behaviours, they need external stimuli and motivations to maintain those.

Priming is the organisational psychology term for creating the cues that motivate a person towards a certain behaviour. For example, an individual would put their recycling in a separate bin only as long as their neighbours do the same. Yet if they have *internalised* recycling behaviours, they don't need cues and further priming to continue separating their rubbish. Another external tool that structures experiences and beliefs into a 'system' is *social regulation*—the feedback loop to individual actions. These tools for behaviour fine-tuning co-exist in the build-up of pro-climate behaviours. Internalisation, priming and social regulation need to be performed in continuity; until the social system crystallises where individuals are convinced enough that their neighbours will act in a similar faith and with similar reliability as they do.

The Fascination with Positivity in Social Identity, and the Repertoire of Pro-climate Identities

Social identity is the response to our social roles and social situations. How do we create our social identity? Do we have a say or choice in adopting one? Can social identities be toxic?

Henri Tajfel (see Tajfel & Turner, 1979) famously founded the grounds of social identity theory—he looked how people start identifying with social groups, exaggerating traits of the in-group and out-groups. Born into a Jewish family in Poland, he studied at Sorbonne in Paris during the outbreak of the World War II, joined the army, became a prisoner-of-war and then was working with refugees and re-settlement in the UK. Together with John Turner, Tajfel established the grounds to understanding prejudice, group identity and perception.

Tajfel was one of the first academics to look at how people's values and beliefs shape their perception. Individuals' mental processes organise the stimuli in a storytelling that fits their worldview—in a basic example, a hungry meat eater may mistake a blurred image of a red flower for a piece of steak (Vaughan, no date). This same effect of biased cognition leads to stereotyping according to racial and ethnicity-based prejudices.

Tajfel's contribution to the field of social identity theory lied in his understanding that simply belonging to a group provided enough grounds for the members of this group to dismiss or discriminate the members of a different group (Tajfel, 1970). This was a tectonic shift from the previously influential group-conflict theory that implied that two groups have incompatible goals hence the conflict arises. Tajfel's experience of Holocaust showed him how hatred and prejudice are borne out of a group identity.

Together with his student John Turner, Henri Tajfel established social identity theory (1970) that defined that group identification is a conscious choice—a cognitive act. In this conceptual paradigm, individuals compare themselves not with individuals, but with other groups. For instance, an Italian cook looks down upon British cooks expecting them to be less skilled at their trade. Moreover, social identity theory makes an essential and much-valued aspect of identity—search for positive self-evaluation—the cornerstone for why group members strive to improve the positive image of a group or throw all the mental resources into believing so. Tajfel and later Turner (1975) discussed the relationship between groups as the 'competition for positive identity'. They also introduced a novel concept for the 1970s—of social change. If members of a group do not find social identity of a group positive, they either leave or strive to improve the positive appeal of the group and its distinction from the others (Tajfel & Turner, 1979).

Social identity serves as a self-reference and a cognitive-emotional array of beliefs. It provides orientation in the world. This heavy reliance on group identity becomes an especially potent breeding ground for propaganda ideas in authoritarian societies, where elites try to associate in-group identity with high morality, cultural superiority and other assumed special characteristics to oppose the outside word. Ethnocentrism breeds on the similar in-group bias—it does not arise from the conflict about resources, but from the inherent beliefs about the superiority of one group over the other (Tajfel & Turner, 1979).

However, further critical debates about social identity theory moved away from the focus on conflict and were more curious about the stability of an identity, and of how social identities across groups (family, neighbourhood, gender, profession, ethnicity) co-exist and overlap (Islam, 2014). The 'repertoire of social identities' is a non-dual proposition to study the intersecting social identities one has (Abrams & Hogg, 1988). The salience of each of these memberships is what matters—and this is mostly activated by context.

Ellemers et al. (2002) take the salience debate further and specifies that the range of characteristics used to judge the superiority of one's group over another can differ depending on each group identity we hold. They also reminds us that self-reliance of individuals will grow, and group identification will weaken when group boundaries are permeable. '(I)t is the social context, rather than specific group features, that determines the evaluative flavour of any given group membership' (Ellemers et al., 2002: no page). This presents an interesting challenge for forming pro-environment group identity as the group is vague and unsolidified. The 'evaluative flavour' in terms of pro-climate action identity may mean alignment with ideological movements, commitment to individual or collective action, neighbourhood or country-level initiatives for the environment—what makes it so interesting in the 2020s is the vague and emerging nature of this repertoire of pro-climate identities.

What self-identification and what cognitive 'package' of interpreting the stimuli would be beneficial to climate action? In the UK, for instance, the climate agenda traditionally sits with the left side of the political spectrum, yet it is under the centre-right government of the Conservative Party that the country has committed to net-zero goals in 2050 (Rathi, 2023). This mixed range of climate messages in the political agenda is beneficial to the post-ideological perception of pro-climate social identity, or—conversely—it may encourage the leading political parties to make the climate message more salient and divisive in their promotional efforts.

Pro-climate identity lies at the crossroads of responsible citizenship, individual empathy, social identity of assumed collective responsibility, group identity of parents looking to better the world for their children, or of Western populations empowered to act, or feeling the guilt for reaping the benefits of fossil fuels economy but not the devastating effects of it on the climate.

The poststructuralist critique to Tajfel's social identity theory points to subjectivity as the defining feature of social identity. Peirce (1995, as cited in McNamara, 1997: 561), for instance, emphasises the importance of three components of subjectivity: the fluidity of the subject, subjectivity as a site of struggle and the identities changing over time. McNamara (1997) supports this correction but points to the historical and structural underpinnings of strong group identities, thus aligning the classic and the recent approaches to social identity theory.

McNamara (1997) brings about the example of Hebrew-speaking immigrant families in Australia whose experience was shaped by the in-group and out-group perceptions of their Jewish identity. In a different context, the family members could be distinguished by playing a specific sport or holding a particular job in a society.

Environmental identity is complex—unless one is a devoted environmental activist, there is no single identity that would rely on climate beliefs and climate goals. It is less rooted in historical processes and struggles as the identity embedded in race, ethnicity, religion or political beliefs.

In the popular culture oversaturated by superhero narratives, the pro-climate identity may be borne out of the empowerment to take action before it is late, to assume the role of the saviours of those in less fortunate positions. Following poststructuralist approach to social identity, the pro-climate identification may mean different things to different people:

1. It may range from moral superiority of those literate about climate change over climate deniers, or a trendy social role of preferring climate-conscious brands over traditional ones—thus signalling a middle-class or upper middle-class social identity based on 'ethical consumption'. This emerging narrative is being largely supported by the fashion industry and glossy magazines alike (Denisova, 2021a). It provides a lifestyle avenue for social identity formation, which nonetheless is aligned with class and socioeconomic affordances.
2. Pro-climate identity may also mean radical anti-consumption and joining the collectives that pledge not to buy new items, to recycle

and reduce consumption, as well as engage in restrictive self-challenges of an allegiance to an ascetic lifestyle even the household budget allows for more.
3. Pro-climate identity may also present toxic tropes for the classist discrimination, attributing more knowledge, enthusiasm and agency to the higher classes in the society, and potentially sowing division between the white and blue collars—if the discourse in popular press is left unattended to fuel this sort of divisive narrative.
4. Pro-climate identity can be anxiety-ridden, apocalyptic and detrimental to mental health—yet it can produce a sort of in-group identification for those who take the problem deeply, emotionally and strive to address it, for both the immediate community and the distant others.
5. Pro-climate identity acts as a corporate veneer for a large portion of global business—the multiple scandals around bogus climate offset companies demonstrate that being a part of climate solutions is a desirable social and economic characteristic. It is also a difficult one to obtain without the dramatic overhaul of business models and emissions—hence pro-climate identity may be an umbrella term for distinct corporate hypocrisy.

Immo Fritsche from University of Leipzig and his colleagues (2017) are proposing a SIMPEA—a Social Identity Model of Pro-Environmental Action. Social identity is the bridge between 'I' and 'we'—it is the role we see ourselves in with regard to the social groups that matter to us. Climate crisis—as Fritsche et al. shrewdly point out (2017)—cannot be observed directly at all times, and we largely rely on social institutions such as scientific community, governments, charities and the media to equip us with the information on the matter. This already creates grounds for the collective experience of the climate change—and it would be only logical to pursue collective efforts in tackling this issue. In this light, the division of pro-climate identity into a repertoire of more or less salient actions and beliefs seems the most likely way forward. Whichever domain the pro-climate identity taps into—ideological, sociocultural, socioeconomic—it can offer a variety of manifestations, with the inherent in-group and out-group characterises. Based on the research analyses, it is unlikely that pro-climate social identity can materialise as a standalone distinct group, but it is even more beneficial to the social good if it stays as a scattered palette of associations and beliefs. For now, the social identities that are connected to the concern for climate change are overlapping in a manner of the Olympic rings. The historical roots of pro-climate identity are political and activist yet not as deep and forceful as those of some other identities—which opens avenues for the creation of the modern narratives, using pro-climate identity repertoire as a starting point for further exploration.

Identity Threat, and How the Media Shape a Great Deal of It

No amount of knowledge and science can beat the visceral reaction of identity threat. Even the most emotional or instructive storytelling, whether shaped as a long-form human story or a multi-character multimedia feature, can affect the minds and hearts of the people if they feel like climate action is 'an attack on who they are' (Hayhoe, 2021). As science Professor Katharine Hayhoe (2021) puts it, the physics that humans use in airplanes and fridges is the same science that lies in the core of climate change models—yet some people become very selective in the applied uses of this science, which parts they keep and which ones they reject.

Identity threat is defined as one's perceived allegiance to a group and the distinction between the allies and threats to that group. For climate action, the clashing groups might be farmers and scientists who argue over the allocation of water or the use of fertilisers (Poff et al., 2003). People may be adapting the NIMBY approach—Not In My Back Yard—when it comes to the installation of nuclear power plants or wind turbines in the vicinity of their houses; the locals may find the former threatening or the latter visually unpleasant.

The affiliation to groups can vary in strength and change over time—the climate change discourse in the US is a strong example. It was previously seen as an exclusively Democratic issue, especially with the documentary produced by Al Gore in the 2000s and the climate denialism by the Republican Donald Trump during his Presidency in the 2020s. In mid-2020s, though, the tide is turning as even Republicans are forced to accommodate pro-climate policies and include the issue in the political campaigning (Nilsen, 2023).

Nevertheless, group identity can be very strong—it affects drastically how people respond to knowledge and research (Hornsey & Imani, 2004, as cited in Fielding & Hornsey, 2016). When it is not possible to disentangle group identity from climate awareness and action, the group allegiance can be used in a constructive way. An example of such approach is the study on attitudes and actions (Jang, 2013)—it showed that Americans who were told that most of their compatriots don't preserve energy well were reluctant to save energy to support climate change policies. However, when they were told that Chinese customers are excessive energy consumers, the American respondents were more interested in pro-climate policies and reasonable consumption. The choice of in-group and out-group comparison matters. In another study (Rabinovich et al., 2012, as cited in Fielding & Hornsey, 2016), British respondents rated themselves as more environmentally friendly than the Americans, but lower than the Swedish. The choice of contrast affected the in-group thinking.

Therefore, ethnocentric group identity is a strong influence on climate behaviours (Jang, 2013; Rabinovich et al., 2012, as cited in Fielding & Hornsey, 2016). Putting blame on the whole population is perceived as a threat to the group identity, and people fight back. Emphasis on in-group responsibility is a poor strategy if the population is under impression that 'all pollute' or 'everybody lies' or hold any other thoughts that alleviate of responsibility and give an erroneous impression of what the conformist behaviour comprises. Moreover, highlighting the superior behaviour of the competing group may be a losing communication tactic—the in-group reaction is likely to be defensive (Fielding & Hornsey, 2016).

What is also unhelpful regarding identity threat is stereotyping that frequently occurs between groups. 'We are the defenders of nature, and you are the destroyers of it' narrative (Opotow & Brook, 2003; Fielding & Hornsey, 2016) is common when environmentalists clash with the groups that defend their values, priorities, businesses and ways of living. A typical example is the ranchers who come under the pressure of pro-environmentalist groups to give up their land to the protected species—ranchers see the green groups as inexperienced, arrogant attackers on their land and ways of living. Some urban protests—when they get radical like some examples of Extinction Rebellion in the UK or Italy, including blocking roads or transport links—come at a cost of clashing with the urban commuters who need to get to their place of work or study uninterrupted.

Another example is the 'Gilet jaunes' violent protests in France in 2018 when protesters set cars on fire and looted shops in opposition to

Emmanuel Macron's fuel tax. The 'haves' and 'have nots' in this case become those living in the urban areas and enjoying a developed public transport infrastructure, and those in the rural areas who rely on their cars—plus, arguably, those who jumped on the bandwagon to challenge the establishment disregarding of the cause.

> Though the tax was favored by Parisians, who have access to efficient public transportation, it was seen as a provocation by struggling residents of the country's rural and suburban areas. (Gross, 2018)

Identity threat and in-group mentality is closely related to the curse of imbalanced political communication of the 2020s—the so-called culture wars, matched with mis- and disinformation. In Summer 2023 in the UK, Prime Minister Rishi Sunak posted a message on Twitter from the car of the legendary UK Prime Minister Margaret Thatcher, assuring the voters that he is 'pro-cars'. This definition—although not mentioning climate directly—clearly confronted those who advocate for the reduction in cars, more control over air pollution and keeping strict speed limits. This political message invited the identification of people from various classes, incomes and environmental stances to pick sides—to decide whether they were pro-car or anti-car.

A related concept that adds a cultural layer to social identity theory is that of cultural cognition (Douglas & Wildavsky, 1982, as cited in Fielding & Hornsey, 2016). Simply put, it suggests that people either see the world through individualistic or collective lens. This vaguely relates to the Conservative-Labour divide in the UK but can also signify the filter that people apply to any communication they receive. Worldviews and values are rather static, while identity is more fluid (Fielding & Hornsey, 2016)—something that media professionals should remember when drafting effective communication.

Among the benefits of the devoted identification with groups is the ability to mobilise for collective action. People with a strong affinity to a chosen group are more willing to put effort in action—especially if they believe that they can win, or when their identity is marginalised ('politicised'—van Zomeren et al., 2010, as cited in Fielding & Hornsey, 2016).

Collective responsibility for climate change and the need for the 'loss and damage' international fund may be more supported by people with collectivist views and rejected by those in favour of individualist values and free market. However, the very same idea can be rejected if the in-group position of, say, working or middle-class people in the country, sees the distribution of funds abroad as a threat to the in-group well-being. In this case, group identity prevails over the worldview.

Visual Power Struggles, or Why Greenpeace Is Loading Glaciers and Bears with Meaning

The visual language of climate change is complicated. Historically, the impeding global crisis was illustrated via the means of polar bears, melting glaciers, windless deserts, as well as the more idyllic reminders of the beauty of the Amazon Forest and species at the brink of extinction. Even some of the most powerful environmental communicators on the planet have been struggling with breathing new life into the tired visual palette of climate messaging. This section examines the achievements and challenges of visualising climate change, and the novel insights in the visual symbols that do not alienate people but bring forward the sense of efficacy related to climate action.

Greenpeace, the legacy environmental organisation, has developed the strategy of challenging the status quo through powerful imagery.

> Greenpeace can be regarded as an organisation with photography as its vital medium... Greenpeace's strategy is implicitly based upon the premise that it can and must counter images of power with forceful images of its own. (Boettger, 2001, p. 12, as cited in Doyle, 2007: 130)

The organisation was a pioneer of the 'mindbomb' concept—when seeking to intervene in the media discourse, Greenpeace activists would distribute the disturbing pictures to the media, in the hope of planting some ideas in the minds of the audience (Denisova, 2019). A famous example arises from the story of a Greenpeace boat chasing a whale trawler in Japan—instead of preventing the hunters from killing the gracious animal, the activists would make striking photos and send them to the major media companies. In an instant, a grim image would decorate the front pages of the newspapers and produce a global impact, rather than stopping one boat from destroying one whale.

Bob Hunter, the mind behind Greenpeace's media stratagem, defined this method as planting a 'mindbomb'—it is a visual metaphor, a disturbing trigger that will get 'activated' when the time comes, for instance, when the audience will get to vote for the green party, or will be inclined to sign a petition against whale hunt. Hunter famously drew inspiration from the ideas of the academic visionary Marshall McLuhan, who postulated that people 'live and think mythically' (Dauvergne & Neville, 2011; Denisova, 2019).

In milder terms, Greenpeace's approach can be seen as photographical 'bearing the witness' to the damage being done to our planet by human action (Doyle, 2007). To explain the rationale behind this choice, Conny Boettger, the former Greenpeace picture editor, calls images the universal language that does not need translation between languages and cultures (Boettger, 2001, as cited in Doyle, 2017: 132).

Nevertheless, Doyle (Doyle, 2017) argues that the strategy of showing the shocking and the beautiful—what has served Greenpeace well since its conception in 1994—may be falling short when it comes to sustainable climate change journalism and activism. Focussed on reporting the breathtaking, creating the symbolic layer for the major developments in the climate change field—whale hunt, oil spills, protests—Greenpeace created its own slow spectacle of climate change discourse.

Although Greenpeace was at the forefront of framing science for popular understanding—they produced a reader's digest to the first 1990 IPCC report—the history of visual climate mythology demonstrates how climate change demands references to other iconography or cultural symbols to reach salience.

The first attempts—the Greenpeace report in the 1990s—invoked nuclear disaster iconography and the title 'climate time bomb' while the next reiteration of climate materials mid-1990s suggested a household-relevant metaphor and a call to action. 'Putting the lid on fossil fuels' title offered a double-act—first, an ancient-civilisation-like allegory of a volcano eruption and, second, a more grounded allusion to an overflowing pot on the stove.

Further down the history lane, the imagery of the Blue Planet photographed by the satellite provided both a symbol of innocence and a testament to human progress (never before space exploration would an image like this be possible). Yet the iconography of the globe in space was later criticised for its detachment from the 'domain of lived experience' (Ingold, 1993, as cited in Doyle, 2007: 139).

The 1997 Greenpeace report relied on the—now familiar—image of a split ice rock. The pristine vast whiteness of the Antarctica landscape served as the reminder of the purity of nature in the face of reckless human action—it was also featuring a tiny human figure off-centre, looking over the cliff. While the contrast between a little human and mammoth nature was striking, alluring to the distant frosty lands was not boosting the relatability factor for many.

The 1998 report focussed on the renewable energy as positive action—it featured the wind turbines on the cover (Doyle, 2007). This shift from problems to the solutions was exacerbated in the next turn in communications—by the early 2000s, Greenpeace set up campaigns against the fossil fuel companies such as BP (British Petroleum) and Esso and directed its rage at the influential elites involved in environmental decision-making. A 2002 campaign advertisement featured a close-up of George Bush-Jr and said 'Don't Buy Esso. Bad company'. This poster was prompted by Esso's backing of Bush's presidential campaign and his subsequent withdrawal from the Kyoto protocol. For the first time, no landscape or symbolic climate imagery was used. The metaphor of 'dirty oil – dirty politics' was deployed instead (Doyle, 2007).

Mid-2000s saw more melting glaciers in Greenpeace's campaign iconography. An experiment with shock campaigning was held too—the NGO released a video in which a family was enjoying themselves on the beach when a plane crashed into a nuclear station on the coast. 'Tell Tony Blair nuclear power is not the answer' (Greenpeace, 2005, as cited in Doyle, 2007: 145). Greenpeace challenged the UK government in court too, arguing against the decision to invest in nuclear power as the means of reducing the use of fossil fuels and tackling climate change.

While the examples above mostly relate to the global, US and UK audiences, there have been localised campaigns too. The work by Greenpeace Canada focussed on polar bear preservation and eulogising the energy efficiency (Slocum, 2004). The many layers of climate communication invoked global and parochial references (Slocum, 2004) in their pursuit of the audience's indignation and commitment to the cause.

Even the ubiquitous polar bear—ubiquitous in climate communication rather than habitat, sadly—is a floating signifier. It means more to some cultures than to the others. For instance, for the Canadians, Innuit and Cree, polar bears are a beloved symbol of the Canadian Arctic, a cultural trigger (Slocum, 2004). The formidable animal used to serve as a source of food of clothing for the First Nations, and is respected and admired ever since, as the symbol of Northerness, which Canadians identify with strongly. Desperate for food and with the natural habitat melting away, the hungry bear has been featured in demonstrations—with protesters dressed as bears—along with posters and media messages. Simultaneously, Greenpeace Canada bet on the bear as the totem of climate campaign due to its universal appeal (Slocum, 2004). Despite such a strong symbolism of bears inherent to the Canadian origin myth, the media exposure and elaborate campaigning, the audience awareness of the climate change in Canada, after the bear story was deployed, still ranked at 5–10% (Slocum, 2004).

In the Indonesian context, Greenpeace Indonesia posted three main themes on their Instagram in late 2010s—the imagery of floods and humans affected, the call to switch to the renewable energy and the explainer and argument against the Omnibus Law (Pramana et al., 2021). The latter is the controversial initiative of the Indonesian authorities that reduces bureaucracy and encourages business initiatives but weakens the environmental protection. Activists argue that that it allows coal mines to operate for 30 years and extend the license every 10 years until the coal runs out (Pramana et al., 2021). The arresting images of the horrible effects of climate change—such as floods and damaged houses—received two or three times more likes than those about renewables and legislation. One of the more successful Instagram posts against coal mining (theme number three) involved optimistic news that a major investor in energy decided not so support coal mining any longer. While the likes on Instagram do not directly translate into climate advocacy, they nonetheless point to the issues people related to the most—disaster striking close to home along with and the images that give the sense of efficacy, indications of the solutions and the feeling of work in progress.

In journalism, the studies show that the visual elements of climate stories matter. In a US-based generalised study of news articles on climate change and audience perception, a range of images and texts were identified that improved the feeling of efficacy among readers (Hart & Feldman, 2016). In an array of images available—from bad flood, smoke arising from a power plant to a climate march, or solar panels—the participants responded the best to the stories that offered visual solutions. The members of the audience reacted positively to the photographs of solar panels and the discussion on active measures to address climate change—this type of communication made them feel empowered. The stories that used the imagery of climate protests, floods or power plants did not generate the feeling of empowerment. Among the reported behaviour changes, the participants identified efforts in energy conservation and the willingness to affect political decisions related to climate change. While the study (Hart & Feldman, 2016) is clear about its limitations—e.g. the story may have overshadowed the poignant imagery—it nonetheless demonstrates how symbols and elements of storytelling can have various effects on the perceived efficacy of the audience members.

In conclusion, facts are material objects that travel (Dumit, 1997, as cited in Slocum, 2004: 417). They serve as the grounding points of the ephemeral discourse, as the power grid for the transmission of ideas about climate change. And yet—**it is just as likely for someone to be passionate about the electrical grid as it is for them to be infatuated with climate facts**. These factual anchors of climate awareness are necessary and indispensable, yet more is needed to make the issue tactile, sensory and close at hand.

The visual array of Greenpeace has been arguably limited in the 1990–2000s (Doyle, 2007), as not all impact of climate change can be seen or visualised in a photograph powerful enough to capture public imagination. The evolution of the influential organisation's campaigning demonstrates the challenges of creating an appealing, not tired, or detached iconography that can pull at the heartstrings and call to action. It has also suffered from the increased desensitisation of the publics to the imagery of the melting glaciers, miserable polar bears and the mesmerising but uninvolving shots of the planet as a whole, looking almost like a souvenir, an emoji, something too big or too timeless to control. The same challenge exists in journalism—while holding the public remit to report truthfully and utilising the real imagery of the real events, the media face a challenge in presenting a non-banal image of a problem or a solution, and find enough visual impact in the limited visual stock of those. The examination of climate imagery in journalism and climate activism calls for much further research. In an oversaturated media economy, images are essential to engage the readers and provide the codes for climate awareness and action that are accessible and inspiring.

Bibliography

Abrams, D., & Hogg, M. A. (1988). Comments on the motivational status of self-esteem in social identity and intergroup discrimination. *European Journal of Social Psychology, 18*(4), 317–334.

Ajzen, I. (1991). The theory of planned behavior. *Organizational Behavior and Human Decision Processes, 50*(2), 179–211.

Allport, F. H. (1924). The group fallacy in relation to social science. *American Journal of Sociology, 29*(6), 688–706.

Allport, G. W. (1962). The general and the unique in psychological science. *Journal of Personality, 30*(3).

Andre, P., Boneva, T., Chopra, F., & Falk, A. (2021). *Fighting climate change: The role of norms, preferences, and moral values.* University of Bonn and University of Cologne, Reinhard Selten Institute (RSI), Bonn and Cologne.

Andreoni, J. (1990). Impure altruism and donations to public goods: A theory of warm glow giving. *The Economic Journal, 100*, 464–477.

Andreoni, J. (2001). The economic of philanthropy. In N. J. Smelser & P. Baltes (Eds.), *International encyclopedia of the social and behavioral sciences*. Elsevier.

Asch, S. E. (1951). Effects of group pressure upon the modification and distortion of judgments. *Groups, leadership, and men* (pp. 177–190).

Baumeister, R. F., & Leary, M. R. (1995). The need to belong: Desire for interpersonal attachments as a fundamental human motivation. *Psychological Bulletin, 117*(3), 497–529. https://doi.org/10.1037/0033-2909.117.3.497

Berger, J., & Milkman, K. L. (2012). What makes online content viral? *Journal of Marketing Research, 49*(2), 192–205.

Boon-Falleur, M., Grandin, A., Baumard, N., & Chevallier, C. (2022). Leveraging social cognition to promote effective climate change mitigation. *Nature Climate Change, 12*(4), 332–338.

Centola, D., Becker, J., Brackbill, D., & Baronchelli, A. (2018). Experimental evidence for tipping points in social convention. *Science, 360*(6393), 1116–1119.

Cialdini, R. B. (2001). The science of persuasion. *Scientific American, 284*(2), 76–81.

Cialdini, R. B., & Schroeder, D. A. (1976). Increasing compliance by legitimizing paltry contributions: When even a penny helps. *Journal of Personality and Social Psychology, 34*(4), 599.

Cialdini, R. B., Wosinska, W., Barrett, D. W., Butner, J., & Gornik-Durose, M. (1999). Compliance with a request in two cultures: The differential influence of social proof and commitment/consistency on collectivists and individualists. *Personality and Social Psychology Bulletin, 25*(10), 1242–1253.

Cialdini, R. B. (2007). *Influence: The psychology of persuasion*. Collins.

Coleman, J. (1988). Social capital in the creation of human capital. *American Journal of Sociology, 94*, 95–120.

Czarniawska, B. (2006). A golden braid: Allport, Goffman, Weick. *Organization Studies, 27*(11), 1661–1674.

Dauvergne, P., & Neville, K. J. (2011). Mindbombs of right and wrong: Cycles of contention in the activist campaign to stop Canada's seal hunt. *Environmental Politics, 20*(2), 192–209.

Denisova, A. (2019). *Internet memes and society: Social, cultural, and political contexts*. Routledge.
Denisova, A. (2021a). *Fashion media and sustainability*. University of Westminster Press.
Denisova, A. (2021b). 'Viral journalism', is it a thing? Adapting quality reporting to shifting social media algorithms and wavering audiences. In *The Routledge companion to political journalism* (pp. 271–278). Routledge.
Denisova, A., (2023). Viral journalism. Strategy, tactics and limitations of the fast spread of content on social media: Case study of the United Kingdom quality publications. *Journalism, 24*(9), 1919–1937.
Doyle, J. (2007). Picturing the clima (c) tic: Greenpeace and the representational politics of climate change communication. *Science as Culture, 16*(2), 129–150.
Dunlap, R. E., & York, R. (2008). The globalization of environmental concern and the limits of the postmaterialist values explanation: Evidence from four multinational surveys. *The Sociological Quarterly, 49*(3), 529–563.
Ellemers, N., Spears, R., & Doosje, B. (2002). Self and social identity. *Annual Review of Psychology, 53*(1), 161–186.
Festinger, L. (1954). A theory of social comparison processes. *Human Relations, 7*, 117–140.
Fielding, K. S., & Hornsey, M. J. (2016). A social identity analysis of climate change and environmental attitudes and behaviors: Insights and opportunities. *Frontiers in Psychology, 7*, 121.
Fischbacher, U., Gächter, S., & Fehr, E. (2001). Are people conditionally cooperative? Evidence from a public goods experiment. *Economics Letters, 71*(3), 397–404.
Fritsche, I., Barth, M., Jugert, P., Masson, T., & Reese, G. (2017). A social identity model of pro-environmental action (SIMPEA). *Psychological Review, 125*(2).
Fuller, R. G., & Sheehy-Skeffington, A. (1974). Effects of group laughter on responses to humorous material: A replication and extension. *Psychological Reports, 35*(1, Pt 2), 531–534.
Goffman, E. (1974). *Frame analysis: An essay on the organization of experience*. Harvard University Press.
Gregson, N., & Crewe, L. (2003). *Second-hand cultures*. Bloomsbury Publishing.
Gross, N. (2018). Is environmentalism just for rich people? *The New York Times*, 14 December. https://www.nytimes.com/2018/12/14/opinion/sunday/yellow-vest-protests-climate.html
Hayhoe, K. (2021). *Saving us: A climate scientist's case for hope and healing in a divided world*. Simon and Schuster.
Hornstein, H. A., Fisch, E., & Holmes, M. (1968). Influence of a model's feeling about his behavior and his relevance as a comparison other on observers' helping behavior. *Journal of Personality and Social Psychology, 10*(3), 222.

Islam, G. (2014). Social identity theory. *Journal of Personality and Social Psychology, 67*(1), 741–763.

Jang, S. M. (2013). Framing responsibility in climate change discourse: Ethnocentric attribution bias, perceived causes, and policy attitudes. *Journal of Environmental Psychology, 36,* 27–36.

Judge, M., Kashima, Y., Steg, L., & Dietz, T. (2023). Environmental decision-making in times of polarization. *Annual Review of Environment and Resources, 48.*

Kahneman, D., & Knetsch, J. L. (1992). Valuing public goods: The purchase of moral satisfaction. *Journal of Environmental Economics and Management, 22*(1), 57–70.

Katz, B., & Kahn, R. L. (1966). *The social psychology of organizations.* Wiley.

Kay, A. C., & Friesen, J. (2011). On social stability and social change: Understanding when system justification does and does not occur. *Current Directions in Psychological Science, 20*(6), 360–364.

Laclau, E., & Mouffe, C. (1985). *Hegemony and socialist strategy: Towards a radical democratic politics.* Verso.

Latane, B., & Darley, J. (1970). *The unresponsive bystander: Why doesn't he help?* Appleton-Century-Crofts.

Latour, B. (2005). Gabriel Tarde and the end of the social. In P. Joyce (Ed.), *The social in question* (pp. 117–132). Routledge.

Lee, M. S., & Ahn, C. S. Y. (2016). Anti-consumption, materialism, and consumer well-being. *Journal of Consumer Affairs, 50*(1), 18–47.

Levy, M. (2021). Depop: the fashion reselling platform used by a third of the UK's 16–24-year olds. And me. I'm 41. *iNews.* 2 February. https://inews.co.uk/news/long-reads/depop-fashion-reselling-platform-upcycling-feature-854799

Loschelder, D. D., Siepelmeyer, H., Fischer, D., & Rubel, J. A. (2019). Dynamic norms drive sustainable consumption: Norm-based nudging helps café customers to avoid disposable to-go-cups. *Journal of Economic Psychology, 75,* 102146.

McNamara, T. (1997). Theorizing social identity; what do we mean by social identity? Competing frameworks, competing discourses. *TESOL Quarterly, 31*(3), 561–567.

Mildenberger, M., & Leiserowitz, A. (2017). Public opinion on climate change: Is there an economy–environment tradeoff? *Environmental Politics, 26*(5), 801–824.

Naeem, M. (2021). Do social media platforms develop consumer panic buying during the fear of Covid-19 pandemic. *Journal of Retailing and Consumer Services, 58,* 102226.

Nguyen, N. (2021). Case study: How Red Bull used rubbish bins to achieve marketing success. *Smith Brothers Media,* 7 July. https://smithbrothersmedia.

com.au/get-smarter/case-study-how-red-bull-used-rubbish-bins-to-achieve-marketing-success/#:~:text=Red%20Bull%20placed%20empty%20cans,based%20on%20its%20'popularity'

Nilsen, E. (2023). Why Republicans can't get out of their climate bind, even as extreme heat overwhelms the US. *CNN*, 30 July. https://edition.cnn.com/2023/07/30/politics/republicans-climate-solutions-heat-wave/index.html

Omarjee, L. (2023). *To make climate change resonate with audiences, connect it to heritage and culture*. Reuters Institute at Oxford University, 4 August. https://reutersinstitute.politics.ox.ac.uk/news/make-climate-change-resonate-audiences-connect-it-heritage-and-culture

ONS. (2022). *Worries about climate change*. Great Britain: September to October. https://www.ons.gov.uk/peoplepopulationandcommunity/wellbeing/articles/worriesaboutclimatechangegreatbritain/septembertooctober2022

Opotow, S., & Brook, A. (2003). Identity and exclusion in rangeland conflict. In S. Clayton & S. Opotow (Eds.), *Identity and the natural environment*. The MIT Press.

Poff, N. L., Allan, J. D., Palmer, M. A., Hart, D. D., Richter, B. D., Arthington, A. H., Rogers, K. H., Meyer, J. L., & Stanford, J. A. (2003). River flows and water wars: Emerging science for environmental decision making. *Frontiers in Ecology and the Environment, 1*(6), 298–306.

Pramana, P. D., Utari, P., & Naini, A. M. I. (2021, April). Symbolic convergence of# ClimateCrisis: A content analysis of Greenpeace Indonesia campaign on Instagram. In *IOP Conference Series: Earth and Environmental Science* (Vol. 724, No. 1, p. 012101). IOP Publishing.

Prentice, D. A., & Miller, D. T. (1993). Pluralistic ignorance and alcohol use on campus: Some consequences of misperceiving the social norm. *Journal of Personality and Social Psychology, 64*(2), 243.

Putnam, R. D. (1993). *Making democracy work: Civic traditions in modern Italy*. Princeton University Press.

Rathi, A. (2023). *Climate capitalism: Winning the race to zero emissions and solving the crisis of our age*. Greystone Books Ltd.

Reingen, P. H. (1982). Test of a list procedure for inducing compliance with a request to donate money. *Journal of Applied Psychology, 67*(1), 110.

Rowlands, I. H. (1995). *The politics of global atmospheric change* (Vol. 1). Manchester University Press.

Schwartz, S. H. (1973). Normative explanations of helping behavior: A critique, proposal, and empirical test. *Journal of Experimental Social Psychology, 9*(4), 349–364.

Schwartz, S. H. (1977). Normative influences on altruism. In *Advances in experimental social psychology* (Vol. 10, pp. 221–279). Academic Press.

Sharot, T. (2011). The optimism bias. *Current Biology, 21*(23), 941–945.

Shearman, S. M., & Yoo, J. H. (2007). "Even a penny will help!": Legitimization of paltry donation and social proof in soliciting donation to a charitable organization. *Communication Research Reports, 24*(4), 271–282.

Slocum, R. (2004). Polar bears and energy-efficient lightbulbs: Strategies to bring climate change home. *Environment and Planning d: Society and Space, 22*(3), 413–438.

Snow, D. A., & Benford, R. D. (1992). Master frames and cycles of protest. *Frontiers in Social Movement Theory, 133*, 155.

Snow, D. A., Rochford Jr, E. B., Worden, S. K., & Benford, R. D. (1986). Frame alignment processes, micromobilization, and movement participation. *American sociological review*, pp. 464–481.

Sparkman, G., & Walton, G. M. (2017). Dynamic norms promote sustainable behavior, even if it is counternormative. *Psychological Science, 28*, 1663–1674.

Steffen, A. (2017). Second-hand consumption as a lifestyle choice. In *International conference on consumer research (ICCR)* (pp. 189–207). DEU.

Stern, P. C., Dietz, T., Abel, T., Guagnano, G. A., & Kalof, L. (1999). A value-belief-norm theory of support for social movements: The case of environmentalism. *Human ecology review*, pp. 81–97.

Tajfel, H. (1970). Experiments in intergroup discrimination. *Scientific American, 223*, 96–102.

Tajfel, H., & Turner, J. C. (1979). An integrative theory of inter-group conflict. In W. G. Austin & S. Worchel (Eds.), *The social psychology of inter-group relations* (pp. 33–47). Brooks/Cole.

Thöni, C., & Volk, S. (2018). Conditional cooperation: Review and refinement. *Economics Letters, 171*, 37–40.

Titmuss, R. (1970). *The gift relationship: From human blood to social policy*. Vintage Books.

Turner, J. C. (1975). Social comparison and social identity: Some prospects for intergroup behaviour. *European Journal of Social Psychology, 5*, 5–34.

Ungar, S. (2000). Knowledge, ignorance and the popular culture: Climate change versus the ozone hole. *Public Understanding of Science, 9*(3), 297.

Vaughan, M. (no date). Henri Tajfel, Britannica. https://www.britannica.com/biography/Henri-Tajfel

W.P. Carey News (2007). *The gentle science of persuasion, part four: Consistency*. W.P. Carey School of Business, Arizona State University, 17 January. https://news.wpcarey.asu.edu/20070117-gentle-science-persuasion-part-four-consistency

Walton, G. M., & Wilson, T. D. (2018). Wise interventions: Psychological remedies for social and personal problems. *Psychological Review, 125*(5), 617.

Weber, K., & Glynn, M. A. (2006). Making sense with institutions: Context, thought and action in Karl Weick's theory. *Organization Studies, 27*(11), 1639–1660.

Weick, K. E. (1979). *The social psychology of organizing* (2nd ed.). Addison-Wesley.

Weyant, J. M. (1984). Applying social psychology to induce charitable donations 1. *Journal of Applied Social Psychology, 14*(5), 441–447.

Wicker, A. W. (1969). Attitudes versus actions: The relationship of verbal and overt behavioral responses to attitude objects. *Journal of Social Issues, 25*(4), 41–78.

Williams, C. C., & Windebank, J. (2002). The 'excluded consumer': A neglected aspect of social exclusion? *Policy & Politics, 30*(4), 501–513.

Zhang, L., Ruiz-Menjivar, J., Luo, B., Liang, Z., & Swisher, M. E. (2020). Predicting climate change mitigation and adaptation behaviors in agricultural production: A comparison of the theory of planned behavior and the Value-Belief-Norm Theory. *Journal of Environmental Psychology, 68*, 101408.

CHAPTER 7

Climate Optimism or Climate Pessimism? Self-Efficacy Boosters and Storytelling for Change

GET YOUR JAB, OR HOW JUST ONE TWEAK OF COMMUNICATION MAKES PEOPLE ACT

In the 1960s, a social psychologist Howard Leventhal and his colleagues conducted an experiment at the University of Yale to see the impact of storytelling and fear on decision-making. They asked a group of students to read the booklet about the risks of the disease called tetanus (Leventhal et al., 1965). The participants were then asked to fill out the questionnaire about their motivation to get vaccinated against tetanus. No other conversation was taking place with the research team.

There were four variants of the leaflet. One with the 'high-fear' description of the disease and a case study of a patient who has it; the one with the 'low-fear' descriptions and a case study; a version with specific recommendations; and a version without recommendations. The researchers detected a high level of arousal in those who have engaged with the most dramatic version of the text. Students experienced anxiety, fear, even nausea. Motivated by fear, participants reported a higher degree of intention to get inoculated. This correlated with the Leventhal et al. (1965) expectation that fear would act as a catalyst for action. Furthermore, it was revealing that some reactions—like nausea—were outlined by the researchers as a potential inhibitor of action. This correlates with the more recent studies of psychological triggers (Berger & Milkman, 2012) that explain that sadness, or, in this case, nausea as a visceral reaction to fear and distress, may stop people from taking action. If one feels utterly disempowered, depressed, nauseated by the scale of the climate crisis, they are likely to disengage from the climate coverage and be confused about the steps worth taking.

In the most illuminating outcomes of the tetanus leaflet experiment, Leventhal et al. (1965) discovered that providing the participants with specific recommendations helped them to reduce the effects of distress and take steps to protect themselves from the disease. These findings are deeply useful for the understanding of climate communication.

Being concerned creates a whirlpool of unpleasant emotions. However, the concerns can be turned into action when matched with **specific recommendations** (in case of the Yale study, the researchers added a map of how to arrive to the vaccination centre from the room where the experiment was taking place). Without specific recommendations, the effects of the high arousal by fear rapidly evaporated after the experiment. When Leventhal and his team called the vaccination centre to inquire on the uptake of the shots by the students in the control groups they discovered a satisfying result. 9 out of 59 (15%) of the participants ended up going to the vaccination centre round the corner and receiving the free tetanus jab. Some of them were in the high-fear group with specific recommendations, others in the low-fear group with specific recommendations, and only one person got vaccinated without having been given specific recommendations. This influential study demonstrates that emotional arousal is useful but not decisive in behavioural change. Clear guidance on the practical steps is crucial.

The finding that fear influences attitude change but not behaviour change has been later confirmed by Leventhal et al. (1966). Examining the sex of the participants of a similar study that used the variables of fearful storytelling and specific/general instructions, they discovered a similar correlation. Neither fear nor practical steps alone were enough to trigger a change in behaviour. In fact, those more vulnerable to the disease (the research used the tetanus example again) and specifically male participants who were vulnerable to tetanus, reported lower motivation to get inoculated. This made the researchers conclude that a highly dramatic, fear-inducing communication combined with specific instructions may cause strong reactions—either the motivation to get immunised, or a resistance. Why would resistance emerge? Leventhal and colleagues suggested the 'coercive' atmosphere (Leventhal et al., 1966: 398), the feeling of pressure and having no choice as the factors triggering resistance, especially among male participants.

Eco-anxiety and Turning Strong Emotions in the Currency of Change

Eco-anxiety, the attitude of intense sense of worry related to climate change, is experienced by many people around the globe. Some of the strongest distress has been detected in younger generation. In one of the largest surveys of the 2020s, Caroline Hickman and her colleagues surveyed 10,000 children and young people (aged 16–25) across the world. The study took place in ten countries: Australia, Brazil, Finland, France, India, Nigeria, Philippines, Portugal, the UK and the US (Hickman et al., 2021). The majority of the respondents declared to be worried about the climate change (59% very or extremely worried; 84% moderately worried).

Relevant to the studies of fear and sadness, more than half of all children and young people surveyed reported the emotions of anger, sadness, anxiety, powerlessness, helplessness and guilt. Among these emotions, anger and anxiety (Berger & Milkman, 2012) are arguably more likely to trigger actions, while others are the deactivating, depressing feelings. The nature of anxiety is important—while generally considered an emotion that encourages action, it can get inhibited by other debilitating emotions, when experienced simultaneously, and thus can be a deactivator, too.

Eco-anxiety has a profound effect on everyday lives. Firstly, it can influence attitudes to everyday tasks and productivity—45% reported that their daily life was affected by the thoughts about climate change, and 75% found the future frightening (Hickman et al., 2021). Here, the presence of fear is clearly linked with the abovementioned depressing emotions and thoughts. Secondly, climate anxiety can worsen when individuals feel a sense of responsibility without clear avenues for taking action on that responsibility, leading to an overwhelming burden. Climate anxiety can therefore be also defined as the uneasy feeling of worry often combined with fear and even guilt about the impact of climate change on the planet (Hickman et al., 2021). Albrecht (2011) is often cited as the creator of the term 'climate anxiety', while reiterations of the definition are many. In a nutshell, climate anxiety is seen as a chronic stress response to the perceived dangers of the climate crisis.

Hickman, an experienced psychotherapist and a researcher at the University of Bath, talks about the persistent worry that climate crisis and the communication about it are causing to children and young adults. She suggests opening the conversation about climate worries with children and adolescents, not only with adults. She argues for the approach of rationalising the discourse, unpacking the worry and addressing the parts of it one by one. '(T)hinking about it, naming it, feeling it, reframing it, relating to it, listening to "it" and renaming it' are all healthy tactics to navigating a complicated conversation,' Hickman (2020: 412) proposes.

Alternatively, there are calls to reduce the conversation about climate change with children. Australian ex-Prime Minster Scott Morrison famously criticised the renowned climate activist Greta Thunberg and other youth climate voices for provoking 'needless anxiety in our children' instead of letting 'kids be kids' (as cited in Hickman, 2020: 413).

However, a more prominent view among academics and media thinkers is the criticism to the tone of climate communication aimed at young people (Ritchie, 2021, 2022), rather than the existence of this communication per se. Some leading figures within Extinction Rebellion, the radical climate movement—such as the founder Roger Hallam or spokesperson Robert Read—have been accused of catastrophising the situation, promoting the message of despair and doom.

> People probably sometimes ask you "What are you going to be when you grow up? But we've reached a point in human history where the question also has to be asked "What are you going to do **if** you grow up?"—Rupert Read, the spokesperson of Extinction Rebellion, as cited in Ritchie (2022: no page)

The scale of climate crisis is hard to grasp—and even harder it may be for some people to reconcile with the human causes of it. As Hickman (2020: 414) devastatingly puts it, 'in relation to the climate crisis we seem to be simultaneously powerfully causative and powerlessly helpless'.

The media coverage of climate change can be useful for informing, educating and potentially empowering the population. However, the persistence of negative news values and the abundance of coverage at a rapid pace may feel like 'repeat traumatic news being delivered that we never quite have time to recover from before the next worrying stories appear' (Hickman, 2020: 415). Neuroscientist Tali Sharot (2011) agrees with that, reminding that fear and anxiety lead to the withdrawal and freeze, checking out instead of leaning in.

Too much fear helps no one. Alarmingly, fear is a potent potion, which can be used by climate deniers to convince people NOT to act. As Hayhoe (2021) cautions, climate sceptics may infer the fear of 'losing our freedoms, eating steaks, driving cars and going on holidays' as a motivation to doubt climate science and stay passive. Fear can inhibit urges to action, but it can also inhibit the openness to climate science.

What about those already convinced and rather overwhelmed by fear? What about the purer, younger ones who take climate communication even more to the heart? The feeling of despair about climate change is common among young people—but can be tweaked into positive behaviour change. Hickman (2020) proposes 'eco-empathy' as the way to appreciate one's feeling of being connected, ability to care and relate to the problems of others. Eco-empathy is suggested as a more rewarding emotion linked to the sense of belonging, compassion and unity. Moreover, when things get overwhelming, individuals need the clear instructions for action to convert the tension into helpful release. Echoing Leventhal et al. (1965), this way out of the arousal and fear must be paved out with high-specificity suggestions and clear paths.

All in all, eco-anxiety is a healthy response (Hickman, 2020) to the complex issue that is climate crisis; yet it can turn inhibiting to the daily tasks and climate action when left unsupported by additional emotional responses and cognitive reasoning. In this regard, the aroused feelings are justified and can be compared to the worry people felt when reading the booklet about tetanus in the study of Leventhal et al. (1965). The scale of the climate crisis is undoubtedly different yet the natural response of alertness, unease, worry is similar to that of the tetanus experiment. The media have the power to divert this energy of distress into meaningful action. Psychotherapy calls for 'naming it', 'listening to it', but also to finding the empowerment within it, identifying paths for achievement and acting upon them as the means to escape the eco-anxiety whirlpool.

THE ROUTE TO ECO-EMPOWERMENT—SELF-EFFICACY

'Missing cat' is a commonplace poster often seen on the corners of urban dwellings. We all have seen this kind of a placard with a gorgeous fluffy feline attached to the posts in your area. Would not your heart sink thinking of the worries of the owners, and the dangers that might befall the poor cat? What usually happens after that sighting is knowing nothing else about the story. Yet in rare cases new posters may emerge, or a scribble on top of the old poster—'Cat found!'. This little detail reminds of the good news that exist in the world, of the care of the strangers, or even simply provides a sense of closure to the worry that some people may experience for the cat and the cat's owners. What is often missing in media coverage is the positive developments of a story, the coverage of the empowering acts of agency—we need to know more about the found 'cats' from the headlines. The constructive narratives that show not just the incidents, but the improvements and developments can help boost the sense of self-efficacy, collective efficacy and motivate people to dare to act.

There is not a one-size-fits-all mechanism that explains how attitudes, emotional triggers and practical information convert into behavioural change. As the work of Leventhal et al. (1965) demonstrates, people's knowledge and their behaviour are not connected by default. They need a link, a connecting matter. This matter can be the practical guidelines for action—but they would not produce any result without the belief in the efficiency of the chosen behaviour.

The existing research on efficacy beliefs derives from the seminal study of the Canadian American psychologist Albert Bandura. Known for his experiments and deep dives into human behaviour and cognitive processes, he has established the concept of self-efficacy. Bandura (1977: 193) postulated that individuals are driven by self-efficacy, which is defined as the sense of being capable of addressing the issues they face with desirable outcomes. People are pursuing the life-long quest of 'creating and strengthening expectations of personal efficacy'. According to Bandura (1977), cognitive capacities and contextual experiences affect the sense of efficacy. When adopting a new behaviour, cognitive processes must be accompanied with social modelling. This means that, even if a person decides to commit to a certain action, they need to be able to imitate the existing behaviours of others, and then put into practice their behavioural change. **Applying one's cognitive decision on practice is the only way to implement the behaviour for good, make it 'second nature'. The next step in solidifying the behavioural change is receiving feedback to one's actions. If the feedback is consistent and clear, a person is more likely to proceed with the newly adopted way of being.**

The cognitive process and modelling do not come without hiccups—for instance, individuals may believe that a behaviour will generate desired outcomes one day, even if for the time being they are being punished for the chosen activity. This idiosyncrasy means that people's beliefs may be dominating their thinking, even overshadowing the response of the world to their actions. The persistence of delusional activity can be seen, for instance, in non-democratic countries, where people may act amorally, if the state or social media propaganda gives them the 'legitimate' excuses to do so (Ressa, 2022). People may ignore immediate feedback from their social circles or be selective in the choice of feedback and give preference to the one that aligns with their current state of beliefs.

For Bandura (1977: 193), efficacy is 'the conviction that one can successfully execute the behaviour required to produce the outcomes'. A belief clouded by doubts does not result in action. When people fear that the situation is beyond their coping skills, they do not act. In the context of climate change, fearmongering and apocalyptic storytelling from the media result in exactly this state of affairs. People may be aware of the crisis but feel unable to act and disempowered to do so. They do not see or realise self-efficacy in climate matters.

Those who apply their self-efficacy beliefs in lightly threatening situations—say, a job interview—and succeed, are likely to gain confidence and improve self-efficacy expectations. Next time, they may be able to do a managerial task or try resolve a work conflict. However, those who withdraw even from the lightly frightening scenarios, due to low self-efficacy beliefs, are preventing themselves from realising their potential, power and coping skills.

The strength of self-efficacy studies has been proven in many contexts, including motivation for health-related behaviours (see Williams & Rhodes, 2016, for an extensive review). Williams (2010), however, corrects the self-efficacy purity by suggesting that when people expect certain outcomes, this can improve the sense of efficacy—as opposed to self-efficacy not being dependent on the expected outcomes, in Bandura's writing. This viewpoint derives from the expectancy-value theory (Atkinson, 1957). This concept posits that people balance off their expectations of the achievements of their actions with the value they put into the task. Various factors affect the commitment to performing a task—including past experiences and social circle's beliefs and values (Eccles et al., 1983). When applied to pro-environmental behaviours, the expectancy-value theory stands strong as long as the context of action facilitates 'green' action. For instance, the ease of access, government policy, a workplace's sustainable efforts can either encourage or hinder pro-climate behaviours (Eccles & Wigfield, 2020). The amalgamation of personal beliefs (including self-efficacy) with situational variables (that boost the validity of expectancy-value theory) are the vigorous factors for the pivot to environmental consciousness and behaviours (Raghu & Rodrigues, 2021).

To enhance the expectations that one's actions are likely to lead to the considerable climate change mitigation, the media come into play. The varied media diets of modern societies nonetheless contribute to the cultural 'structure of feeling', per Raymond Williams's (1983) term. The classic yet underexplored tenet of British cultural studies refers to the implicit, subtly pronounced trends and sentiments that derive from the agenda of the moment, from the drifts in cultural production and lifestyle habits. The 'structure of feeling' is a loose parallel concept to ideology, it allures to the vague arrangement of ideas and sentiments, an intuition, that possess the minds of people at a certain period and forms culture. 'The structure is unsatisfactory, but the feeling is valid', as Simpson (1992: 15) delicately puts it. It is not uncommon to refer to the media—including both professional journalism and the vast Internet spaces—as a significant supplier to the 'structure of feeling').

When applying self-efficacy to climate, one can see that even implementing small behavioural changes can affect positively one's ability to cope with the scale of the crisis. Presenting the stories of small goals and small victories in the media is not trivial—it aids the overall self-efficacy expectations of the individual consuming the media messages and living through the climate anxiety or uncertainty. One can give up meat or improve their recycling habits and already feel that they are contributing to climate action. Removing the presence of climate denialists in the media is another change that can help improve the 'structure of feeling' related to climate mitigation and adaptation validity and necessity—and build climate self-efficacy in the audience. It might be unreasonable to expect individuals to move from no action to massive advocacy campaigns—however, small steps and a continued build-up of self-efficacy expectations are instrumental in addressing climate crisis.

The journey of improving one's self-efficacy expectations should ideally be strengthened by positive experiences. One should put into practice what they believe in, to gain more of this desired sense of mastery that self-efficacy provides. 'Performance accomplishments' is the useful term to measure how successes improve mastery expectations, and how failures lower them (Bandura, 1977). Repeated successes lead to a strong sense of agency, and even failures along the way may enhance a more motivated, challenge-prone behavioural style. This phenomenon explains how people need small achievable tasks and successes to celebrate along the way, if they were to improve their confidence and self-efficacy in the long run.

Similar to a university setting, where students are given relatively easy tasks in the first semester of the first year, and increasingly complex challenges in the second and third year, this progression of goals and successes teaches resilience but also self-efficacy. The caveat lies in the processing of failures. Self-efficacy development does not mean that the road is paved with successes alone—but it is the variation of tasks and challenges that an individual can pursue that allows for a failure at a low cost to motivation. To continue with the university metaphor, a way to mitigate the potential build-up of a failure pattern is to offer a range of varied tasks to the students. This allows them to tap into various talents, skills and motivations to identify the areas of accomplishment and those less likely to lead to a triumph. Moreover, the beauty of self-efficacy establishment lies in its effect on other areas of life—an individual, once empowered in one sphere, applies this feeling of self and one's agency to other activities and endeavours in their life (Bandura, 1977).

The effects of modelling on behavioural change are weaker than implementing cognitive beliefs matched with action. Nonetheless, observing someone struggling with the fear and inaction overcome their issues may be potent for the individuals to follow in the same path. It has been studied (Kazdin, 1974, as cited in Bandura, 1977: 197) that modelling on a successful, advanced behaviour is less efficient than modelling on a person with the similar struggles who is nevertheless showing resilience and new mastery in dealing with their issues. This demonstrates how media campaigns on climate change should feature 'a person like me', not celebrities or otherwise successful people. These may be good as baits for attention but are less relatable and less triggering for a behavioural change. For the same reason, the classic Hollywood stories of an 'underdog', a humble individual working through their limitations and misadventures, to achieve a greater good, success and recognition, are so popular. The media need to provide stories that show the build-up of self-efficacy that are not a straight line, but the stories of trials and errors. These can empower the individuals who have not experienced a linear success in life but are working hard to achieve their goals.

An interesting aspect to note here is that a campaign that presents diversified individuals who—despite coming from all walks of life and backgrounds—can overcome the issue is a positive move. A variety of behavioural models (Kadzin, 1974 and 1975, as cited in Bandura, 1977: 1977) is even more helpful for an individual than a singular model. If people of the most humbled or troubled backgrounds can succeed—an individual is likely to think—then the behavioural change is doable and worth implementing. This scheme may remind one of group therapy for substance abuse, where sharing individual stories is generally seen to be helpful to emphasise the common nature of experiences and highlight the path towards recovery. It shows our shared human nature but also provides a range of relatable models to learn from and be inspired by.[1]

Well-defined outcomes are essential in behaviour modelling. Showing examples of clear action with clear results is of utmost importance for behavioural change. In the case of climate crisis, the paths of mitigating the issue are complex, larger-than-life and ambiguous. Yet still, providing clear explanations to the outcomes of individual efforts is helpful in motivating the climate-friendly behavioural adjustments. A suggestion to reduce one's vast consumption of fast-fashion items must be presented with the clear data on the effects of this activity—how much water, resources, landfill emissions one avoids by decreasing their shopping sprees. Another example of a lifestyle adjustment that can bring a meaningful reduction in one's emissions is flying. Avoiding flying in general or at least refraining from long-haul flights should be socially praised. The media should help individuals see lifestyle changes like these as a tool of (limited, but tangible) control over the complex issue of climate change. Highlighting progress—even a relative one—is instrumental in improving one's self-efficacy (Bandura, 1977).

To check self-efficacy theory, Bandura (1977) conducted an experiment with the people fearing snakes. He applied two approaches to cure their phobia—one group was asked to stay in the room with the boa constrictor, touch it, let it stay in their lap and to open their arms in front

---

[1] Interestingly, the Learned Helplessness theory by Maier and Seligman (1976) postulates that, as a result of facing uncontrollable adverse events, people learn to have no control over things. They stop trying to inflict any chance, even though the areas of control that they have and may influence are vast. This may happen in an unresponsive environment or the environment when one is easily punished. To change this paradigm, a change of the environment is necessary along with the development of individual competencies and achievements.

of its head. The second group was merely observing the therapist perform these tasks with the snake. As a result, more people in the snake-touching group reported the increased sense of self-efficacy—this desired blend of confidence and mastery that made them empowered to face the phobia but also feel more confident in performing other tasks.

Collective efficacy forms a separate branch of knowledge—it evaluates the beliefs in the collective capability to put the effort in and achieve results (Bandura, 1977; Choi & Hart, 2021; Koletsou & Mancy, 2011). When applied to climate change, a high appraisal of collective efficacy may mean the belief that many people are capable of switching off the lights when not using the room. Wording can be important—the European Social Survey (2015, as cited in Choi & Hart, 2021: 4) makes an important distinction between 'able' and 'likely' when designing questionnaires for the public. When scientists try to measure the collective efficacy, they are advised to include the word 'likely' as it indicates the intention to perform an action, not just the physical and psychological capability to do so. Therefore, the studies on collective efficacy are constrained by exploring the intentions rather than actions of the individuals—yet they are valuable in demonstrating the way people evaluate their capacity. Furthermore, collective efficacy is a useful indicator of trust in social fabric, in the idea that other people may act in the same way as an individual, in addressing climate change.

Climate change is the problem that needs a large-scale collective action—and the effects of individual efforts are not immediately apparent. It is a Catch-22 conundrum—the more people believe in collective action and the efficiency of it, the more likely they are to engage with it (Lubell, 2002). Nonetheless, it takes much effort to convince the majority that the tipping point, the critical mass of the climate-conscious people has emerged—as discussed in the previous chapters, 'social proof' is essential to generate the feeling that the majority are on the same page. This is why it is essential that climate communication explains and emphasises the fruit-bearing collective action, shows the results of citizen mobilisation and society-level initiatives to address climate change. The perceptions of citizen efficacy, beliefs in trust and reciprocity largely affect the intention to engage in climate action (Lubell, 2002).

The crossover between self-efficacy and collective efficacy needs a crucial instrument—communication. Private attitudes change faster than the social norms (Miller & McFarland, 1991, as cited in Prentice & Miller, 1993: 2). This is why public discourse—that comprises journalism, social media, brands, promotional communication—helps to speed up the process of establishing and disseminating the new norm, the new status quo. Similarly, in the study of various efficacy types—personal, collective, governmental—Meijers et al. (2023) found that people should believe that their collective actions could produce results. They would not be as motivated for behavioural change if they only believed in their own efficacy, or their own potential in achieving results. In order for the private sphere beliefs to grow into self-efficacy, an individual needs to see some evidence that collective action bears fruit. It is more important than seeing individual action bear fruit. In connection to that, acknowledging the efficiency of the government in dealing with the climate issues can also stimulate the feelings of empowerment and efficacy of public sphere agency (Meijers et al., 2023).

Apocalyptic Storytelling: Why It Awes, but in the Wrong Way

In July 2023, The New York Times (Rubin, 2023) published a mesmerisingly executed and meticulously researched long feature on the effects of climate change on the 'cradle of civilisation',[2] the area of land that once used to be called 'Mesopotamia'. The article explored the devastating effects of global warming on Iraq, Iran, Syria and neighbouring countries—embraced as the biblical 'Garden of Eden', 'the fertile crescent'—widely recognised as one of the birthplaces of civilisation. The wheel and the irrigation system were invented there. Through the interactive layout of the piece, with the panoramic shots of the dry lands and drone footage of headless palm trees, interspersed with sharp pull quotes and moving human stories, the overall message was crystal clear—climate change severely damages the area and forces masses of people to leave or adapt to dire conditions. Aggravated by poor geopolitical decisions—such as the construction of dams that prevent water supply between neighbours and favour national interests—the feature left a sour taste. Many climate journalists and scientists shared and recommended the piece on social media.

---

[2] 'Rubin, A. (2023). A climate warning from the cradle of civilisation. *The New York Times*, 29 July. Available at: https://www.nytimes.com/2023/07/29/world/middleeast/iraq-water-crisis-desertification.html.

The most upvoted comment under the article on The New York Times website (over 1,200 readers recommended it) said:

> Thank you for an outstanding in-depth and beautifully illustrated article. The problems caused by climate change, lack of water and population growth seem both grave and unsurmountable.

'Grave and unsurmountable' were the words that caught my attention—it was a sober and sad indictment to the trouble brought by climate change and humankind's futility in dealing with it. This important journalistic effort created a restrained yet intoxicating metaphor that tied together the beginnings of the human civilisation with its possible end. Knowingly or not, the authors infused their formidable reportage with mythological, religious, apocalyptic undertones that made the reading irresistible but the feeling of hope—almost unreachable.

Does apocalyptic storytelling help to awaken the publics to the realities of climate change? Does it incite the feeling of agency and efficacy? Does viral sharing mean the encouragement to act and make a dent in the dramatic situation?

To answer these questions, another big journalistic success on climate coverage comes to mind. In 2017 New York magazine published the now-viral essay 'The Uninhabitable Earth' by David Wallace-Wells (it ended up being the most-read article in the history of the magazine). The writer focussed on the most extreme forecasts of the impact of climate change—and captured public imagination with his tenebrous, exaggerated, electrifying prose.[3]

These examples of emotionally potent, hypnotising and scary journalism catch the attention of the readers and probably challenge climate indifference. Yet what effect they have on empowerment?

Framing is the media theory that suggests deconstructing the media message through the lens of main actors, reasons and consequences, as well as indicating moral judgement and possible remedies (Entman, 1993). Frames are the commonly recognisable and ritualised structures

---

[3] Five years later, Wallace-Wells published a much more optimistic piece 'Beyond Catastrophe' (The New York Times, 2022) that praised the coordinated global effort and commitments, the decreasing price of renewables and scientific innovation in addressing climate change. He received accusations of shifting gears and being unreasonably optimistic.

of organising information, which emphasise some elements of meaning at the expense of others (Entman, 1993). Framing is important to identify issues, attribute causes and define solutions (Entman, 1993; Foust & O'Shannon Murphy, 2009).

The media framing of climate change in both The New York Times's 'cradle of civilisation' feature and New York magazine's forecast long piece are exemplary of the apocalyptic approach to climate reporting. Two types of catastrophising have been noticed in both elite and popular media outlets in the US: the Armageddon-like portrayal of climate disasters as the humanity's Fate; and the 'comically' catastrophic, assuming that humanity has made its mistakes, but some agency is still left on behalf of humans to mitigate the calamity (Foust & O'Shannon Murphy, 2009). The use of 'comic' here indicates the dichotomy of tragedy-comedy as the archetypal frames of storytelling from the ancient times, not some degree of humour embedded in the situation.

Apocalyptic discourse is enchanting—it overwhelms with awe, and so do the major global religions' visions of the apocalypse. Apocalyptic storytelling on climate change in the media bears its characteristic traits. It offers an array of symbols that need decoding, it calls for reflection and it enables specific individuals as prophets (Foust & O'Shannon Murphy, 2009). In the doomist media stories on climate, scientists become the 'prophets' that turn abstract data into clear signs of the impending doom ('the butterfly population diminishing', 'no ice in the Arctic at least once this summer') (see Ignatius, 2006, as cited in Foust & O'Shannon Murphy, 2009: no page). Scientists connect the dots and explain the links between distant butterflies and the heatwaves in people's towns. They make climate change more sizeable and hold the keys to understanding all the major and minor developments that are hard to grasp by an average human mind. One risk here is that, in populist environments, fake prophets may be plummeted to media fame, side-lining scientists and knowledge holders. There is a high risk that climate deniers, lobbyists, populist politicians, as well as fossil fuel-supported 'scientists' may rise to prominence through the mere means of explaining complex issues simply, suggesting easy solutions and mobilising fearful population against the wrong target. **Apocalyptic framing removes individual and collective agency and thus makes societies susceptible to populism and conspiracy.**

Time is another bewildering aspect of doomist storytelling. Even after the scientific explanations, the climate crisis retains a lot of complication and inconsistency—the timings and precise effects are blurred, which also fits the apocalyptic storytelling like a glove. While not the fault of scientists or journalists per se—the forecasts and climate modelling can be as precise as the data they feed into them—incomprehensible time nonetheless adds to disorientation of the readers.

Similar to the biblical visions of the Apocalypse, humans are not told specifically when and where will the next disaster strike, but it has been made clear to them that it will happen. 'The accelerated time' is a distinct feature of apocalyptic climate coverage—it leaves the reader unsure whether the crisis has already come, or will happen overnight, or in a decade's time (Foust & O'Shannon Murphy, 2009). This compression of time places the events outside of human control, it amplifies the dramatic discourse and reverberates with determinism.

There is a curious tendency to compare current crisis with the catastrophes of the past—volcano eruptions, extinction of the dinosaurs, analogous ancient catastrophes (Foust & O'Shannon Murphy, 2009). This unhelpful metaphor looks good in the headlines but does little to assure humans that they still can turn the tide.

Apocalyptic storytelling may be useful in drawing people in and infusing them with the high dose of fear, but it does little in terms of agency. The 'protection motivation theory' (Rogers, 1983) explains this paradox—when people face the solvable crisis and are assuming that their actions are sufficient to mitigate it, they will be more likely to act. However, if people estimate that the threat is larger than life, they will resist the proposed action. Following from this, presenting audience with the fatalistic narrative is a mistake. If it leads people to believe that they can barely intervene in the unfolding crisis, nothing but low agency and low action will entail (Lorenzoni et al., 2007). Oddly, the cosmic dramaticism alleviates humans of the responsibility for the crisis—it focusses on the grand forces of nature (Foust & O'Shannon Murphy, 2009) and gently removes human agency from the picture.

Related to this, even the names of activist collectives such as Extinction Rebellion bear apocalyptic undertones. While it has been postulated by scientists that even the most dramatic increase in the global temperature won't lead to the actual eradication of the human race, the name of the activist group suggests otherwise. The real fight against climate change is not about keeping human species on the planet but mitigating the effects of their anthropocentric activities that influence flora, fauna and humans alike.

The warning to avoid ultimatums such as 'act in ten years or face an apocalypse' (Foust & O'Shannon Murphy, 2009) is echoed by the head of the IPCC in 2023 Jim Skea. Ultimatums may seem a helpful specification of time—yet they also echo the dramatic framing of 'we are doomed'. In the run-up to mind 2020s, the goal of avoiding the 1.5C increase in global temperatures was crowned as the main symbol of climate action—it has been used by activists, politicians and the media. When it became clear that the world has heated up to 1.5 degrees Celsius and not enough action has been taken to reduce the use of fossil fuels, many despaired. This is why Jim Skea, the Chairman of the Intergovernmental Panel on Climate Change in 2023, calls for less obsession with symbols and for the rejection of doomism in media coverage.

> 'If you constantly communicate the message that we are all doomed to extinction, then that paralyses people and prevents them from taking the necessary steps to get a grip on climate change', Skea insists. 'The world won't end if it warms by more than 1.5 degrees. It will, however, be a more dangerous world' (Deutsche-Welle, 2023).

How to minimise the withdrawal effects and the audience's perceived low agency in response to doomist coverage? Scholars advise connecting climate action with the existing discourses (Foust & O'Shannon Murphy, 2009). These can include energy sovereignty, national identification—although the risk here lies in the pitfalls of populism. It may well be too easy for certain right-wing politicians to fall into racism, anti-migration rhetoric and othering those in the countries most exposed to the climate effects.

Sizing down the crisis to the human scale is a much healthier approach than doomism. While climate change is not going away anytime soon, doomwriting and doomscrolling helps no one. As Sarah Jaquette Ray, the chair of environmental studies at California State Polytechnic University, puts it, when the problem is too big, we feel very small. A remedy to that is to take action, look for environmental groups to join, assess what is feasible to do in the realm of a household, community, workplace, voting decisions in a country—while maintaining the interest and eagerness to act. 'If the problem is so big and we're so small, which is what the doom narrative is telling us, then we need to make the problem smaller and us bigger', Ray suggests, while acknowledging that 'It takes courage and discipline to keep cultivating community and health right where you are, especially amid such bad news' (Buckley, 2023).

Is Climate Optimism an Empowering Strategy or a Route to Complacency?

Journalist Liza Featherstone in her article for New Republic pitched climate optimism and climate pessimism against one another—to curious conclusions. 'The Case Against Both Climate Hope and Climate Despair' (2023) argues that climate optimism indulges citizens to continue their lives as if nothing happened, perhaps with an extra item recycled or fewer plane journeys taken. She cites a climate justice writer Mary Annaise Heglar's term 'hopeium' as a warning to of complacency or feeling unreasonably positive about the crisis. Heglar defines 'hopeium' as an unfounded belief that someone else with come up with solutions (Buckley, 2023), the position of waiting for some magic, *deus ex machina* to appear and save us all.

There is truth to both doomism and optimism—doomism raises people's awareness but impedes action. Optimism activates action and hope but may result in weak types of action and overall passivity.

What both the writer Heglar, journalist Featherstone, and the IPCC Chairman Skea agree on is that media professionals should avoid the 'dead ends' in their messages. Ultimatums, obsession with the almost-lost causes (coral reefs, Arctic ice), and phrasing the headlines that act like a hammer and prioritise the gloomier projections are unhelpful to motivate change. As Heglar puts it, there is a difference in lives and livelihoods saved between 1.5 and 1.6 degrees of warming (Buckley, 2023). Despairing when the 1.5C mercurial line is reached does not mean stopping the mitigation and adaptation efforts—just the opposite.

Featherstone (2023: no page) concords: '*We can't stop it, but we can save many human lives, great civilisations, beloved species, and vital ecosystems. Call it a tempered pessimism or a cautious optimism, but it's the only defensible approach to climate change*'.

Both approaches are true—there are reasons for optimism and pessimism. There is less reason now to engage audiences with doomism—they either withdraw from the discourse or dismiss the whole issue as a lost case. The awareness on climate change, its reasons and effects, has been built in many countries around the globe—and this is the best achievement of doomist storytelling. Now it is time to move on, adjust the storytelling to the needs of maintaining an informed and active population. More positive developments must be acknowledged now, to make sure the audiences are given the social proof of other parts of the society working towards climate mitigation, and they need the reassurance that some progress is being made and the case is not lost.

As the Pulitzer Prize-winning New Yorker journalist and climate writer Elizabeth Kolbert (2022) postulates, *'To say that amazing work is being done to combat climate change and to say that almost no progress has been made is not a contradiction; it's a simple statement of fact'*.

How to reconcile a strong emotional effect—fear, despair, worry— with the ability of the readers to act? Does a feature like the New York Times' deeply researched feature on 'Mesopotamia'—sharply highlighting major problems affecting 40+ million Iraqis and their neighbours—help to inspire action or the feeling of doom? Other commentators, while praising the piece, asked for more practical storytelling too, for a larger offering of solutions journalism.

If readers perceive climate storytelling as 'grave and unsurmountable', the effects may be varied—from engagement with climate activism to withdrawal, denial, othering. Let's examine the likelihood of these scenarios of audience reaction and action.

Powerful reportage like the one on modern 'Mesopotamia' can help strengthen personal norms and values but is not likely to be proportional to action. People face several barriers between intention and action (Whitmarsh, 2009)—it is what they do that matters more than what they say. Those actions that are popular are usually linked with financial savings, health, or personal preferences. Recycling, for example, is an easier pro-environmental act than sacrificing a car or reducing energy use (Ibid.). 'Environmental citizenship' is often proposed as the norm-based policy intervention for pro-climate action (Dobson, 2003; Whitmarsh, 2009). This concept aligns rights with responsibilities and calls for fostering moral-based motivations within communities, educating citizens of their duties and presenting these as a social norm.

Cognitive dissonance is an obstacle to this (Whitmarsh, 2009)—people may overestimate their climate action and underestimate their carbon footprint. Other potent factors at play are the differences between one society and another—what works for Sweden may not work for the UK, what is fine for Switzerland and Germany with high levels of trust in the politicians and the media, will not be efficient in, say, Venezuela, Iraq, China with other ideas of governance, institutions, trust, collective or individualistic priorities among the citizens.

Yet what is more likely to happen within the readers as a reaction to the formidably reported Mesopotamia feature in the New York Times is withdrawal and sadness. High arousal emotions—fear, anxiety, anger, guilt—make individuals more likely to talk or act, while low-arousal emotions—sadness, powerlessness, confusion—are known as deactivating (Feldman Barrett & Russell, 1998; Leviston et al., 2014). Consuming devastating stories about climate change may result in the symptoms of anxiety, powerlessness, sadness—what some psychology researchers (see Gifford & Gifford, 2016) call 'pre-traumatic stress disorder'. In those communities directly affected by the drastic effects of the climate change the symptoms become more severe, as people witness the 'story' unfold in front of them, affecting jobs, habitats and livelihoods. In these cases, the post-traumatic stress disorder (PTSD) becomes a reality. Women, children, elderly, the economically disadvantaged and first responders are more likely to develop strong distress reaction to the events—most people would recover over time, but some might develop a dysfunction (Dodgen et al., 2016).

Those with existing mental health issues are more likely to feel deeply upset by the disconcerting climate change coverage (Gifford & Gifford, 2016). The low trust in the others and the society is another factor undermining resilience and ability to cope with climate-related stress and stressful communication (Dodgen et al., 2016).

To mitigate the reaction of withdrawal and distress, Gifford and Gifford (2016: 295) propose media messages that 'acknowledge strong emotions, provide a local context, encourage collective action, and use appropriate imagery'. Other helpful tools are spiritual growth and promoting a strong sense of community. In general, taking an active position to tackle climate change has been known as the healthy, positive response that acts as a buffer to distress (Dodgen et al., 2016). It does not mean no distress is experienced—but it paves the way to draw plans and actions for adaptation and mitigation, take at least some aspects of the global crisis under control. This approach aligns with Bandura's self-efficacy and the resistance to populism and doomism—it enhances the sense of efficacy experienced on individual and collective levels and maintains the audience's engagement with the information on the issue and curiosity in solutions; it protects the democratic powers of individuals in affecting the wider society and defends them from fake prophets and the abuse of power.

WHAT ANTI-DRUNK DRIVING CAMPAIGNS CAN TEACH CLIMATE COMMUNICATORS?

The message is clear—don't drink and drive, you are risking lives. Yet the need to say it again and again shows the contentious human trait—inconsistency, and the need to be encouraged to pursue the virtuous behaviour and relinquish the vicious one. This section delves into the studies on campaigning that can be helpful for the communicators on climate change, both in journalism and persuasive media.

Anti-drunk driving campaigners have to come up with new creative ideas *every year*. Why so? Drunk driving significantly affects reaction, judgement, vision, information processing (Hanningan et al., 1999, as cited in Cismaru et al., 2009), making accidents, injury and fatality much more likely. Yet drivers still need regular reminders not to risk their and other people's lives.

Some of the more successful campaigns—as revealed by a study spanning the US, UK, Australia, Canada and New Zealand—took into account age, gender and the context of drinking. Others pointed to the localised consequences, instead of the more general warnings. Research shows that the media is not a change agent, but an assisting tool in the gradual restructuring of public's idea of the issues related to drunk driving (Bang, 2000, as cited in Cismaru et al., 2009). Mass media campaigns, combined with journalism attention, are effective in reducing the incidents of drunk driving.

The campaigns that are effective appeal to the feeling of vulnerability and to the severity of the risks. In other words, if a person can relate to the risks of drunk driving, if they feel vulnerable and afraid (for their own or other people's lives), they will be more likely to change their behaviour (Thesenvitz, 2000, as cited in Cismaru et al., 2009). If a person does not feel vulnerable, of if they don't find the threat severe enough, they won't act.

How does it relate to climate change? The more people feel vulnerable to the effects of climate change (from immediate heatwaves to the impact on habitats, elderly, migration surge, conflicts that come along with the more unliveable conditions), the more incited they will be to act. For these reasons, campaigns that seek to involve people in climate activism, social or political behaviour change, need to decide which threat to focus on—and identify what makes the target audience of the campaign feel vulnerable. This would vary from country to country, age, gender, background, context to context. Which risk to nominate as the leading one—this decision should come from the audience research and understanding which particular threats are on top of people's agenda, where they see more severe threats to their particular livelihoods and those of their immediate kin, community and country.

One big weakness of the current climate change campaigns—and climate communication in general—is the lack of efficacy instructions. How exactly can humans, who feel so small in the face of a colossal challenge, can act and contribute to the transformation? In the case of drunk driving, the intended action is clear—do not drink and drive—yet in the case of climate change the range of available actions remains blurry.

This is where the challenge lies—if people do not feel capable of addressing the issue they are presented with, they remain frustrated and less inclined to act. Response efficacy and self-efficacy have been identified as major turning points in campaign efficiency—they increase both adaptive intentions and behaviours (Becheur & Das, 2018; Cismaru et al., 2009).

For example, a US campaign 'Tie One On For Safety' has been running for 20 years. Launched by the respected organisation Mothers Against Drink Driving (MADD), it provides the gruesome statistics on drunk-driving fatalities and then invites people to tie a silver ribbon on their vehicle as a commitment to sober driving. It then distributes leaflets around particular times of the year when drunk driving is more likely to occur—i.e. 'Each year, nationally, more than 1,000 people typically die between Thanksgiving through New Year's in drunk driving crashes' (Cismaru et al., 2009). Another campaign called Designate a Driver—also from MADD—suggests nominating a specific sober person within each group at a party who will be responsible for driving the others home. It adds a note that this person will not be the 'least drunk' but should actually only consume non-alcoholic beverages—otherwise the groups should keep some money aside for the taxi fare. A campaign likes this addresses the many conditions for efficiency—it demonstrates vulnerability and severe threat, and provides clear and accessible avenues for action, while also gratifying sober drivers for contributing to the great community cause. The sliver ribbon on the car acts as a reminder of the pledge and adds a visible virtue to the owner's community reputation.

In the UK context, the THINK! Christmas Drink Drive Campaign took a different approach—it chose a different idea of the threat usually associated with drunk driving. It specifically catered the approach to the male 17–29 drivers. Acknowledging personal motivations and fears of a particular group in the population became a potent strategy. After rounds of qualitative and quantitative research with the 17–29 male drivers in the UK, the campaign creators identified the most resonating threat—that of drunk-driving conviction—that became the central focus of the campaign. TV, radio, online advertising and advertising in pubs sought to remind young men that their life can be ruined if they drank, drove and triggered an accident. The campaigners found that the threat of a car crash was perceived with less fear and relatability than personal consequences for this specific group (Cismaru et al., 2009). The campaign focussed on vulnerability and severity. It did not provide specific instructions on self-efficacy, yet the main message that it was seeking to cultivate—do not drink and drive otherwise you are very likely to ruin your life—came through clearly.

The creative approach of a campaign has a significant effect on memorability—those with fewer words are easier to recollect and repeat to the others (Pešić et al., 2023). From Serbia (Pešić et al., 2023) to Ethiopia (Negi et al., 2020), anti-drunk-driving campaigns demonstrate efficiency in increasing awareness of the effects of alcohol on the driver's systems, and create affective visions of vulnerability and threats, especially among young people. What is interesting is that sometimes the campaigns make young people think twice when boarding the car of a family member or friend under alcohol influence. This creates an additional level of peer pressure within families and groups—another big aspect to help reduce vicious behaviours. In general, low-fear campaigns are aimed at improving knowledge, while high-fear campaigns seek to change behaviours (Yadav & Kobayashi, 2015).

The channel and time of communication matter also—primetime campaign slots bring more effective reduction in drunk driving than daytime and night-time television (Niederdeppe et al., 2017). However, young people who often are at the centre of anti-drunk-driving campaigns, rarely watch broadcast television nowadays. Social media channels, outdoor (transport, streets) and indoor advertising (pubs) play a prominent role. Research on social media use of anti-drunk-driving campaigns is limited (Yadav & Kobayashi, 2015). Social media initiatives often make part of larger campaigns that include legal interventions, checks of drivers' breath, leaflets, advertisements and messages in other media formats.

Journalism, curiously, plays a significant role in assisting the campaigns' visibility and advocacy for the change in policy (Yanovitzky, 2002). It has been discovered that users on social media are likely to share news headlines about drunk-driving-related deaths (Shaw, 2020). The consistent heightened media attention to driving under substance is a low-cost tool to keep the momentum around the reduction in drunk driving (Yanovitzky, 2002).

What these studies on anti-drunk-driving communication mean for climate change messaging is the need to come up with creative solutions again and again—as the message sinks in, people get accustomed to it, and then governments, NGOs and media have to start again from scratch to add novelty to the message.

Another lesson from these studies is the need for accurate and comprehensive research of the very specific target groups that climate communication seeks to capture. **Two main aspects must be identified—the vulnerability and severity of threats to each particular social group**. These may not be obvious—for older people the climate change threats might be linked to health, or it might be the concern over migration, or about their children and their prospects; for younger people, the threats might be related to jobs and health (Creech et al., 1999), or the conservation of natural habitat and preservation of species. As discussed previously, group identity plays a significant role in establishing the priorities and risks for a particular group—and these have to be examined through qualitative and quantitative methods prior to designing the campaign.

An additional layer that increases the campaign efficiency is the instructions, or examples of action. Humans need clarity on how to act—this increases their engagement and sense of capability of dealing with the crisis. In anti-drunk-driving campaign, a non-act (do not drink and drive) and act (nominate a designated driver or put money aside for the taxi) are the most prominent instructions. In climate change communication, non-acts (no flying, no car, not eating meat, not buying new clothes—or at least a reduction in these) can be one kind of message. Calling for acts (signing petitions, taking part in protests, changing a gas boiler for a heat pump, switching to an electric car, choosing vegetarian diet) is a different strategy. In both cases, the priorities and aspirations, vulnerabilities and threats of the specific target population need to be examined in full—with appropriate and feasible instructions provided. Peer pressure or community values are of help, too—increasing the perception that the majority in the community are doing the right thing is a strong motivator for action among the others.

Shaming Me, Shaming You?

'Shame is a soul-eating emotion', the psychoanalyst Carl Gustav Jung postulated.

The feeling of inadequacy, of being observed by others as a lesser being, or one with the socially punishable behaviours, choices, looks—shame is a self-conscious feeling that is not pleasant to experience. A close relative of shame is guilt—the feeling that a person has done something bad to the others and induced their suffering through their words, action or inaction. Both shame and guilt are self-conscious in a sense that they are concentrated on the role of self in a social context, they position self as the reason for negative impressions or consequences. These feelings may be experienced in public and private, relate to moral transgressions (Tracy & Robins, 2006, as cited in Agrawal & Duhachek, 2010). They are neither compatible with the ideal self nor are they sustainable over time. An individual strives to shake these feelings off as they are negative and contribute to negative self-esteem (Agrawal & Duhachek, 2010).

Media messaging that appeals to shame and guilt can be productive as long as it offers paths to mitigation or release from these feelings; not aggravating the existing sentiments. Excellent research on excessive drinking (Agrawal & Duhachek, 2010) has demonstrated how shame can be induced through positioning others as observers, while guilt is induced by positioning others as victims of an individual's exaggerated alcohol consumption. What has also been discovered is that, when a person already feels guilty or shameful of their drinking, the pressure from the anti-drinking campaign may put them on the defensive mode (Agrawal & Duhachek, 2010).

This encounter between the existing emotions and the emotions generated by the campaign are called 'compatible' or 'incompatible' appeal (Agrawal & Duhachek, 2010). The effectiveness of the compatible appeal is lower than the incompatible one. With the compatible appeal, people with the existing shame around their drinking may hypothesise that others—not them—are creating lots of issues for the people around them and are seen unfavourably. It is a biased defensive psychological mechanism that seeks to avoid additional shame on top of the existing one.

Incompatible appeal is more efficient in triggering self-reflection—it refers to the situations when people are taken aback by the new feelings, new self-awareness, a surprising understanding of the reputational damage and worry inflicted by excessive drinking and the social risks that come with it.

Climate communication in the Global North has evident guilt-inducing notes about it. The research (Rathi, 2023) shows that the few in the rich countries are responsible for the greenhouse gas emissions that will cost dearly to the millions in the poorer weather-exposed locations. The science is direct and legitimate. Yet is it the best media strategy to incentivise people to act based on guilt? The findings on compatibility appeal demonstrate that it is a flawed approach—when repeated extensively or when processed by the recipients prone to shame and guilt, these appeals will not lead to a positive action (Tangney & Dearing, 2003). This communication would rather start an elaborate process of inner psychological reasoning that would result in 'mood repair' for the recipient and essentially brushing away the additional shame or guilt on offer (Agrawal & Duhachek, 2010; Becheur & Das, 2018). This psychological mechanism of defensive coping results in low efficacy of the campaigning messages.

The news media in developed countries are reporting the consensus on the responsibility of the richer economics for emissions and also for climate mitigations (Post et al., 2019). However, they also convey slightly different messaging depending on the perceived political efficacy of the country. If a country is capably managed, the media discourse on climate mitigation and adaptation—and to a certain extent, guilt—is more tolerated. When it comes to the media discourse in developing countries, until 2009, many emerging economies were reluctant to attribute too much responsibility—and power—to the developed world (Post et al., 2019). At that point in time, emerging economies were keen to present themselves as equal partners in international climate negotiations. However, with the rising global consensus on the supranational effects of emissions, the developing countries shifted the responsibility to the likes of the UN and other international influential organisations, as well as developed countries (Post et al., 2019). This narrative—albeit based on the scientific and economic data—needs further research as the effects of it on populations in the Global North might be undesired when infused with the frame of shame or guilt for the developed countries. It might be more productive to seek the incompatible appeal, when shame is attributed to the incidents, industries, developments, rather than countries and populations as a whole. Following this narrative, the audiences in the Global North will have the opportunity to perceive themselves as the 'good ones', the people willing to make a dent in the global problem, rather than bearing the weight of responsibility for the many tons of emissions. A healthier approach, psychologically, is to find the constructive ways to mitigate the effects of climate change, while allowing the populations of the developed countries some narrative that is not tainted by shame or guilt. Otherwise, the defensive mechanisms of reasoning might result in withdrawal or anger towards the cause.

The Hero Narratives, or How Humans Like to Crown Their Idols and Then See Them Fall

'Be a HERO! Flush only toilet paper, not sanitary products or nappies down the drain!'—a cheerful, brightly coloured note screamed in a loo of a train taking passengers from Glasgow to London. The reminder to do the right thing—whether it is using the toilet correctly to avoid the repairment costs and a long wait for hundreds of passengers, or preventing drunk driving (as just discussed), or making a difference in a flawed society—all these appeals often utilise one of the oldest tropes on the planet, the Hero narrative. It is not limited to the actions of the old or the young, to the serious or trivial matters—it is ubiquitous in human communication. For example, if you are a primary school child in the UK, most likely your local library has invited you to become 'a reading hero' and competitively devour a couple of classic book titles while on school holidays.

What is it in the human longing to act or feel like a hero—even in the most mundane of activities? The hero storytelling arch has been incredibly powerful over centuries of human storytelling. 'The Hero with a Thousand Faces' (2008), a seminal tome by Joseph Campbell, suggested that all human narrative—from religion to fiction—boils down to the Hero protagonist who reluctantly responds to a call, overcomes his and the world's flaws and saves the day, emerging personally transformed as a result.

The Hero's journey (also known as the Monomyth) has been explored from many sides—it allows for a simpler, dichotomous vision of the world where people are either heroes or villains. It celebrates personal growth and discovery thus making a potent link to the mental health discourse and identity construction. Although the initial conceptualisation by Campbell was criticised for omitting female protagonists and reinforcing heteronormative patriarchal values (Bond & Christensen, 2021), the Monomyth's legacy in popular imagination stands strong.

Among the more discreet layers of meaning in the Hero's Journey is that it delivers a deeply humanistic message—religious or not—that there is an ounce of goodness even in the worst members of the society. There is a chance to opt for better decisions, right the wrong, save the day through sacrifice and spiritual or moral reincarnation. Superheroes have enjoyed an unprecedented, century-long domination in the globalised popular culture (Ndalianis, 2007). Since the 1930, both male and female supernatural characters escaped the confines of comic books and graphic novels and turned into the money machines for blockbuster movies and a source of solace for the populations battered by various crises.

The persistence of the superhero narrative is often connected with the crisis in global identity that was felt particularly acutely after the 9/11 attacks on the World Trade Center in New York (Ndalianis, 2007). Since then, a myriad of 'professional heroes' (Ndalianis, 2007), such as police detectives, doctors, lawyers, talk show hosts and sometimes investigative reporters, have occupied the cultural consciousness through television series, films and reality shows. The 2010s brought the influencers into this realm, celebrating their constructed uniqueness and carefully curated 'authenticity'.

The characteristics of trending heroes and superheroes reveal a lot about the ideological and sociocultural myths of the society where these characters proliferate (Ndalianis, 2007). From this perspective, it is curious to examine the emergence of the climate (super)heroes in the likes of the Swedish teenager Greta Thunberg and the British documentary maker and biologist Sir David Attenborough, the author of the blockbuster BBC nature documentaries *Blue Planet I* and *Blue Planet II*. An impassionate child and a wise man in his 90s make for an odd duo of the globally recognised climate change action advocates. There are indigenous activists and there are pro-climate stakeholders and celebrities, yet arguably none of them reached the level of climate stardom of Thunberg and Attenborough.

It might have been the media fascination with determination (in Thunberg's case) and competence (in Attenborough) that skyrocketed the two unlikely heroes to absolute fame. Thunberg's commitment to *Fridays for Future* (a school strike that she was doing every Friday instead of attending classes in order to bring attention to the climate change) piqued the interest of the Scandinavian and then international media; this was further amplified by her eloquence and clarity of ideas that she conveyed in front of many world leaders. In case of Sir David Attenborough, people were awed by the supreme high-quality production values of his *Blue Planet II*—an anthropomorphic storytelling where miniscule details from the lives of the most unreachable and spectacular species on Earth were presented in high resolution; their struggles and joys were shown through the lens of human relatability—as if foxes and penguins were society members with their household duties and family dynamics.

Heroes perform many healthy functions in a society. Both Greta Thunberg and Sir David Attenborough exhibit the classic traits of heroes with a capital H—they celebrate some of the best human qualities. People are drawn to these bright, selfless, charismatic individuals who can motivate and lead (Allison & Cecilione, 2016). Heroes are useful in media campaigns and storytelling—they ground the problem or phenomenon in a particular individual. Humans have an inherent need for stories. Stories allow to process the complexity of the world, find guidance and relatability, heal the psychological wounds, or release the unhelpful emotions—even the war veterans with the post-traumatic stress disorder find it helpful to write a personal story of survival and recovery (Green & Brock, 2005, as cited in Allison & Cecilione, 2016: 5).

Then, heroes are democratic devices—the unfading success of Superhero movies demonstrates the accessibility of the narrative. People engage with the Hero narrative as it is free from elitism and does not require an advanced literacy. It also holds promise of self-reinvention and empowerment. Gotham City, the fictional hometown of Batman, is famously inspired by New York, the hub of creativity, diverse communities and opportunities for many (Reynolds, 1994), demonstrating a fascinating interlacing of truth and fiction.

The Hero narrative is often open to individual interpretation. It absorbs and refers to the previous Hero storylines in history. Thus, Greta Thunberg is compared to the likes of Joan of Arc, the fifteenth-the century French political icon, and Pippi Longstocking, the imaginative, witty and strong protagonist of children literature in Sweden in the 1960–1970s (Etherington, 2023; Natov, 2007). While dramatically different, these two cultural references resonate with the public as selective interpretations of Thunberg's herodom and what it entails—a selfless fight against the fossil fuels-domination, or the rise to power of a clever young female lead who is claiming her place and agency in the world dominated by the patriarchal tradition.

Moreover, hero stories improve our emotional intelligence—this is why children learn about the social dynamics, positive and dark feelings through fairy tales. The folklore stories are full of encounters with witches, elves, gnomes, evil spirit and neglect—arguably even the gloomiest fairy tales out there, those by the Grimm brothers, 'add clarity and salience' to dramatic emotions and give a greater sense of meaning and purpose (Bettelheim, 1976, as cited in Allison & Cecilione, 2016: 5).

The Hero narrative is…interactive—it is so schematic that it invites collaboration. Heroes are co-created in the media. Similarly to how people may project their desires on popular actors or fictional characters—attribute them the qualities they may not have or never show—the mediated heroes are borne out of complex co-creation. Greta Thunberg's public persona is bred by her social media posts, public appearances and discourses, journalistic articles and comments, opinions online, politicians and stakeholders' speeches who invoke her name for virtuous or villainous purposes (Brugger & Wieser, 2022).

There are several issues with the lone hero narrative. Firstly, while Greta Thunberg's existence and rise to influence present a persuasive role model for young people, she comes under a lot of pressure, expectations and scrutiny. The Hero narrative demands exceptionalism of the protagonist (Brugger & Wieser, 2022)—it is often accompanied by the theme of The Chosen One. It implies that the individual might have been selected by the Destiny or transcendental powers and therefore is most likely to overcome the mammoth challenges that lie ahead of them. From such fictional characters as Neo in *The Matrix*, or *Die Hard's* Bruce Willis's policeman, to the talented football players and coaches, there is a myriad of ways to be a Hero yet they all boil down to certain exceptionalism that has to be proven again and again; it is not static and relies on constant confirmation. The Hero narrative is surprisingly self-contradictory—it seems like anyone can become one, and yet only a few can truly rise up to the occasion.

Secondly, the Hero narrative expects a higher level of competence, skills, tenacity than average—and is often accompanied with the expectation of natural leadership skills. Some of the mediated 'heroes' within their respected professional fields are actress Meryl Streep, ex-basketball player Michael Jordan (Allison and Goethals, 2011), or a football coach Pep Guardiola. Although there is no logical explanation why individual's own brilliance should come alongside excellent skills of managing the brilliance of the others, the Hero narrative constructs—metaphorically speaking—the most ambitious job vacancy description in the world. If the Heroes get it right—and they're capable of inspiring and rewarding the others, then the community is in a bliss. The presence of competent and wise leaders—heroes—gives those around them the sense of 'elevation', i.e. calm resolution to do better and evolve into a better version of self (Haidt, 2003, as cited in Allison & Cecilione, 2016: 8). Yet if the leaders are not managerially gifted, the effects on the community are destructive—the more venerated are the leaders, the Heroes, the more miserable will be the disappointment of those whom they lead.

Thirdly, the simplified narrative of Heroes discounts complexity and nuance. Political scientists note patterns of populism—such as 'truth-telling'—in Greta Thunberg's communication style (Nordensvard & Ketola, 2022). She operates in the space marked by uncertainty and fear—climate change is complex—hence why the clarity and conviction of her talk is attractive to many. From the political communication perspective, Thunberg takes on the role of a hero with a clearly defined villain—the fossil fuels-powered society that refuses to overhaul its energy politics and lifestyle (Ibid.). The iconography is powerful too—the 'underdog' is a neurodivergent girl, while the elites are usually middle-aged male global leaders wearing corporate suits. This contrast is hugely symbolic and evokes the storyline of David versus Goliath.

Fourthly, we do not choose heroes—they choose us (Allison & Cecilione, 2016). The attractiveness of Heroes is driven by the psychological archetypes, which Carl Jung narrows down to, ultimately, Heroes and Villains. The theory of archetypes, as brought to prominence by Carl Jung, the Swiss psychiatrist, suggests that the humankind carries an inherited compendium of themes, personality traits, relationship patterns embedded in the depth of our sub-consciousness. This means that certain traits and characteristics are bound to find popularity—whether the person is virtuous or not, is another story. From this deeply settled enchantment with Heroes derives another trait of Hero worship—the hidden desire to dethrone them, punish any less-than-perfect behaviour, and this is why many Heroes are likely to fall from grace. This dark side of hero Narrative also opens room for interpretation and speculation when it comes to climate heroes—the vilification of Greta Thunberg is intense and vile. It comes from all kinds of emotional responses to her activism—from envy and fear to burnout and confusion.

The fifth issue with the Hero/Superhero narrative is that, in classic Hollywood rendition of Superhero-hood, the protagonist preserves the society from a threat—not trying to revolutionise it or introduce a better way of life (Reynolds, 1994). This means that Heroes are more useful with short-term goals rather than long-term crises such as climate change.

The last but not least issue with the Hero narrative is that it neglects the power of collective action and the more nuanced understanding of the many ways how other individuals—with less charisma, visibility and resources—can make a dent in the problem (Moriarty, 2021). This is why a children literature expert Roni Natov (2007) proposes adding the storyline called 'community as a hero'. In addition to the classical tropes of protagonist-focussed storytelling, she suggests including more narratives of group collaboration, unity for the common goals, and feeling and being a part of the community. A Hero in this case is allowed to both identify heavily with the community but also act separately from it, thus respecting individual agency and common good all at once.

In summary, the Hero storytelling trope is incredibly strong and appealing for the human beings across centuries. It has empowered and fascinated people for generations—and the current climate action Heroes bear the characteristics of the classic protagonists of the past. They are selfless, determined, charismatic and motivate for awareness and action. However, as any Hero trope, it risks bearing the dark tones and attacks on the Hero—the lawed human nature likes to throne its idols as much as see them fall, and this aspect of Hero narrative is not to be neglected. To mitigate the effects of high expectations and increased pressure on the nominated climate Heroes, it is strongly advised that the media focus on inaugurating the community initiatives—whether social collectives, industries or other enterprises where people come together for the sake of addressing climate change—as a scaffolding to the fragile Monomyth, and an alternative to the lone Hero trope.

The Powerful Acts of Storytelling—A Tool for Intervention and Climate Empowerment

How to find the happy middle ground between the seriousness of climate science and data and the accessibility of storytelling? The proposed format to achieve the effect of clarity and empowerment of the audience is 'the powerful acts of storytelling'. It refers to a media artefact that relies on strong data, original reporting, serves as a standalone impactful piece and utilises engaging production techniques that may include multimedia and variety of neutral and ironic writing.

Another term for these exceptionally appealing media outputs is 'view from the moon'. In my research on viral journalism (2023), the deputy editor for The Economist Tom Standage referred to some of the most-shared stories of his publication as 'the view from the moon'. These stories are a crossover of news and features—they discover new big things and provide an unexpected perspective on a specific topic from a specific country. This format may be seen in opposition to national-focussed or 'parochial' approach of many national titles. The 'view of the moon' is the stories that generate awe aplenty and make people intrigued and educated. Some of the Economist's most-shared stories on social media included a video on criminal groups and religion in El-Salvador (580,000 views, 500 shares on Facebook); online grocery shop Ocado's Artificial Intelligence and air traffic technology (700,000 views, 5,800 shares on Facebook). 'A view from the moon' is a striking feature that looks into a curious topic and supplies an in-depth analysis of it.

Noteworthy is the fact that the 'view from the moon' discoveries of The Economist featured no clickbait, animals or children—the 'usual suspects' of online virality. '(P)eople like our analysis of big global themes', asserts Standage (as cited in Denisova, 2023).

'View from the moon' is far from interpretive journalism, a phenomenon of infusing journalism pieces with contextualisation, explanation and sometimes even speculation of the author (Brüggemann & Engesser, 2017; Fahy & Nisbet, 2011). It is more of a fresh thinking on the matter. What story has not been covered in relation to a specific community—e.g. how would British fish and chips shops struggle with the diminishing fishing catch? A mix of two sparkly, trending topics can yield curious returns—global warming and dormant diseases in the permafrost, or mosquitos travelling up North and bringing trouble along with them.

One such story is the long feature by the journalist Jeff Goodell in Rolling Stone, which has received Covering Climate Now 2021 award for some of the best journalism on the matter. A long feature called 'How Climate Change Is Ushering in a New Pandemic Era' features cinematically dramatic, visceral scene descriptions, interviews with the victims of insect bites that were close to becoming life-threatening, and quotes from the scientists at the forefront of new disease prevention; it is sprinkled with wit and resourceful storytelling.

It starts with a story of an American woman called Jennifer who was bitten by a mosquito. It was not an ordinary insect bite but the one that carried a tropical disease called dengue fever. Jennifer had a terrible five days of fever but was lucky to recover. What the long journalism feature in Rolling Stone tells the readers about is the incredible interconnection between humans, flora and fauna, and the multiple dangers of climate change that throw things, like a distorted jigsaw, in unexpected and random directions. Jeff Goodell, a journalist who wrote the long feature 'How Climate Change Is Ushering in a New Pandemic Era' (2020) in that Rolling Stone piece, demonstrated the mastery of storytelling—he moved from the most graphic (or cinematographic even) descriptions of the mosquitos biting humans, to interviewing scientists and explaining their findings in lay terms. He breezily used the romantic comedy term 'meet cute' to describe the turning point of disease mutations, such as the Covid-19 origin 'myth' and the virus passing from a bat to a human to put the world on hold in 2020–21. The skilful mix of colour prose, mind-blowing facts, expert comments and heart-wrenching dialogues with the members of disadvantaged communities in Florida who will suffer the most from the effects of climate change—as well as hints towards how the citizens and the country might act to mitigate all of this—made for a powerful act of storytelling. In fact, research shows that precisely these acts of media coverage—focussed, accessible, engaging, bringing the best tools of journalism storytelling to the issue—are very potent in changing attitudes on climate change action.

On a similar note, powerful documentaries such as *Seaspiracy* that explored the damaging domino-like effects of overfishing and rising temperature levels can be another punchy 'view from the moon' format to connect the previously loosely disconnected dots. **Documentaries are overtaking news organisations in their impact on the audiences around the world**, confirms the annual report by Reuters Institute at

Oxford University (Robertson, 2022). People find compelling storytelling, matched with the striking visuals and emotional pulls, much more memorable than traditional media coverage. This trend is visible across countries—from Portugal, Greece, Australia to Norway, between 30 and 55% of respondents prefer documentaries to the media coverage. Only in Japan and Chile there is more interest in the news organisations' stories on climate change, while in the US the documentaries are enjoying the same amount of attention as routine media coverage (Robertson, 2022).

The 'view from the moon' criteria—novelty, fresh angle, a curious learning curve, low ideological bias, unexpected connections—can help overcome climate fatigue among the audience. They serve as potent attention boosters and leave a dramatic impression on the population. They can stand out in the abundance of info-noise, get shared on social media and are accessible to people from all ideological camps. While this prospect obviously sounds like a dream, the truth is that this kind of high-quality reporting and features are usually limited to the higher-end publications that can afford hiring experienced reporters, writers, sub-editors and whole teams willing to experiment with multimedia and interactive storytelling.

Curiosity and fresh approach—coming back to the drawing board of journalism and approaching climate storytelling not as a no-click dead-zone but a chance to explore the beauty and the quirks of humanity—can have a better chance at capturing the audience's imagination than high numbers of doom-and-gloom news updates on the matter.

Bibliography

Agrawal, N., & Duhachek, A. (2010). Emotional compatibility and the effectiveness of antidrinking messages: A defensive processing perspective on shame and guilt. *Journal of Marketing Research, 47*(2), 263–273. https://doi.org/10.1509/jmkr.47.2.263

Albrecht, G. (2011). Chronic environmental change: Emerging 'psychoterratic' syndromes. In I. Weissbecker (Ed.), *Climate change and human well-being: Global challenges and opportunities* (pp. 43–56). Springer.

Allison, S. T., & Goethals, G. R. (2011). *Heroes: What they do and why we need them*. Oxford University Press.

Allison, S., & Cecilione, J. (2016). Paradoxical truths in heroic leadership: Implications for leadership development and effectiveness. In *Leadership paradoxes* (pp. 73–92). Routledge.

Atkinson, J. W. (1957). Motivational determinants of risk-taking behavior. *Psychological review, 64*(6p1), 359.

Bandura, A. (1977). Self-efficacy: Toward a unifying theory of behavioral change. *Psychological Review, 84*(2), 191.

Becheur, I., & Das, A. (2018). From elicitation to persuasion: Assessing the structure and effectiveness of differential emotions in anti-drunk-driving campaigns. *Journal of Promotion Management, 24*(1), 83–102.

Berger, J., & Milkman, K. L. (2012). What makes online content viral? *Journal of Marketing Research, 49*(2), 192–205.

Bond, S. E., & Christensen, J. (2021). *The man behind the myth: Should we question the hero's journey?* Los Angeles Review of Books, essay, August 21. https://lareviewofbooks.org/article/the-man-behind-the-myth-should-we-question-the-heros-journey/

Brüggemann, M., & Engesser, S. (2017). Beyond false balance: How interpretive journalism shapes media coverage of climate change. *Global Environmental Change, 42,* 58–67.

Brugger, T., & Wieser, V. E. (2022). The iconization of Greta Thunberg: the role of myths in co-creating a person brand. Chapter 25. In *Research Handbook on Brand Co-Creation: Theory, Practice and Ethical Implications.*

Buckley, C. (2023). 'OK Doomer' and the climate advocates who say it's not too late. *The New York Times,* 23 March. https://www.nytimes.com/2022/03/22/climate/climate-change-ok-doomer.html

Campbell, J. (2008). *The hero with a thousand faces* (Vol. 17). New World Library.

Choi, S., & Hart, P. S. (2021). The influence of different efficacy constructs on energy conservation intentions and climate change policy support. *Journal of Environmental Psychology, 75,* 101618.

Cismaru, M., Lavack, A. M., & Markewich, E. (2009). Social marketing campaigns aimed at preventing drunk driving: A review and recommendations. *International Marketing Review, 26*(3), 292–311.

Creech, H., Buckler, C., Innes, L., & Larochelle, S. (1999). A youth strategy for public outreach on climate change. *International Institute for Sustainable Development Business Trust, Manitoba.*

Denisova, A. (2023). 'Viral journalism', is it a thing? Adapting quality reporting to shifting social media algorithms and wavering audiences. In *The Routledge companion to political journalism* (pp. 271–278). Routledge.

Deutsche-Welle. (2023). *Don't overstate 1.5 degrees C threat, new IPCC head says,* 30 July. https://www.dw.com/en/climate-change-do-not-overstate-15-degrees-threat/a-66386523

Dobson, A. (2003). *Citizenship and the environment.* OUP Oxford.

Dodgen, D., Donato, D., Kelly, N., La Greca, A., Morganstein, J., Reser, J., Ruzek, J., et al. (2016). Ch. 8: Mental health and well-being. In A. Crimmins,

J. Balbus, J. L. Gamble, C. B. Beard, J. E. Bell, D. Dodgen, R. J. Eisen, N. Fann, M. D. Hawkins, S. C. Herring, L. Jantarasami, D. M. Mills, S. Saha, M. C. Sarofim, J. Trtanj, & L. Ziska (Eds.), *The impacts of climate change on human health in the United States: A scientific assessment* (pp. 217–246). U.S. Global Change Research Program. https://doi.org/10.7930/J0TX3C9H.

Eccles, J. S., & Wigfield, A. (2020). From expectancy-value theory to situated expectancy-value theory: A developmental, social cognitive, and sociocultural perspective on motivation. *Contemporary Educational Psychology, 61*, 101859.

Eccles, J., Adler, T. F., Futterman, R., Goff, S. B., Kaczala, C. M., Meece, J., & Midgley, C. (1983). Expectancies, values and academic behaviors. In J. T. Spence (Ed.), *Achievement and achievement motives*. W. H. Freeman.

Entman, R. M. (1993). Framing: Toward clarification of a fractured paradigm. *Journal of Communication, 43*(4), 51–58.

Etherington, M. (2023). *Environmental education: An interdisciplinary approach to nature*. Wipf and Stock Publishers.

Fahy, D., & Nisbet, M. C. (2011). The science journalist online: Shifting roles and emerging practices. *Journalism, 12*(7), 778–793.

Featherstone, L. (2023). The case against both climate hope and climate despair. *The New Republic*, 31 July. https://newrepublic.com/article/174719/case-climate-hope-climate-despair

Feldman Barrett, L., & Russell, J. A. (1998). Independence and bipolarity in the structure of current affect. *Journal of Personality and Social Psychology, 74*(4), 967.

Foust, C. R., & O'Shannon Murphy, W. (2009). Revealing and reframing apocalyptic tragedy in global warming discourse. *Environmental Communication, 3*(2), 151–167.

Gifford, E., & Gifford, R. (2016). The largely unacknowledged impact of climate change on mental health. *Bulletin of the Atomic Scientists, 72*(5), 292–297.

Goodell, J. (2020). How climate change is ushering in a new pandemic era. *Rolling Stone*. December 7. https://www.rollingstone.com/culture/culture-features/climate-change-risks-infectious-diseases-covid-19-ebola-dengue-1098923/

Hayhoe, K. (2021). *Saving us: A climate scientist's case for hope and healing in a divided world*. Simon and Schuster.

Hickman, C. (2020). We need to (find a way to) talk about... Eco-anxiety. *Journal of Social Work Practice, 34*(4), 411–424.

Hickman, C., Marks, E., Pihkala, P., Clayton, S., Lewandowski, R. E., Mayall, E. E., Wray, B., Mellor, C., & Van Susteren, L. (2021). Climate anxiety in children and young people and their beliefs about government responses to climate change: A global survey. *The Lancet Planetary Health, 5*(12), 863-e873.

Kolbert, E. (2022). Climate change from A to Z. *New Yorker*, 28 November. https://www.newyorker.com/magazine/2022/11/28/climate-change-from-a-to-z

Koletsou, A., & Mancy, R. (2011). Which efficacy constructs for large-scale social dilemma problems? Individual and collective forms of efficacy and outcome expectancies in the context of climate change mitigation. *Risk Management*, *13*(4), 184–208. https://doi.org/10.1057/rm.2011.12

Leventhal, H., Jones, S., & Trembly, G. (1966). Sex differences in attitude and behavior change under conditions of fear and specific instructions. *Journal of Experimental Social Psychology*, *2*(4), 387–399.

Leventhal, H., Singer, R., & Jones, S. (1965). Effects of fear and specificity of recommendation upon attitudes and behavior. *Journal of Personality and Social Psychology*, *2*(1), 20.

Leviston, Z., Price, J., & Bishop, B. (2014). Imagining climate change: The role of implicit associations and affective psychological distancing in climate change responses. *European Journal of Social Psychology*, *44*(5), 441–454.

Lorenzoni, I., Nicholson-Cole, S., & Whitmarsh, L. (2007). Barriers perceived to engaging with climate change among the UK public and their policy implications. *Global Environmental Change*, *17*(3–4), 445–459.

Lubell, M. (2002). Environmental activism as collective action. *Environment and Behavior*, *34*(4), 431–454.

Maier, S. F., & Seligman, M. E. (1976). Learned helplessness: Theory and evidence. *Journal of Experimental Psychology: General*, *105*(1), 3.

Meijers, M. H., Wonneberger, A., Azrout, R., & Brick, C. (2023). Introducing and testing the personal-collective-governmental efficacy typology: How personal, collective, and governmental efficacy subtypes are associated with differential environmental actions. *Journal of Environmental Psychology*, *85*, 101915.

Moriarty, S. (2021). Modeling environmental heroes in literature for children: Stories of youth climate activist Greta Thunberg. *The Lion and the Unicorn*, *45*(2), 192–210.

Natov, R. (2007). Pippi and Ronia Astrid Lindgren's light and dark pastoral. *Barnboken*, *30*(1–2), 14639.

Ndalianis, A. (2007). Do we need another hero? In W. Haslem, A. Ndalianis, & C. J. Mackie (Eds.), *Super/heroes: From Hercules to superman* (pp. 1–11). New Academia Publishing.

Negi, N. S., Schmidt, K., Morozova, I., Addis, T., Kidane, S., Nigus, A., ... & Murukutla, N. (2020). Effectiveness of a Drinking and Driving Campaign on Knowledge, Attitudes, and Behavior Among Drivers in Addis Ababa. *Frontiers in Sustainable Cities*, *2*, 51.

Niederdeppe, J., Avery, R., & Miller, E. N. (2017). Alcohol-control public service announcements (PSAs) and drunk-driving fatal accidents in the United States, 1996–2010. *Preventive Medicine, 99*, 320–325.

Nordensvard, J., & Ketola, M. (2022). Populism as an act of storytelling: Analyzing the climate change narratives of Donald Trump and Greta Thunberg as populist truth-tellers. *Environmental Politics, 31*(5), 861–882.

Pešić, D., Pešić, D., Trifunović, A., & Čičević, S. (2023). What affects the perception of a drunk driving campaign? *Transportation Research Record, 2677*(5), 196–210.

Post, S., Kleinen-von Königslöw, K., & Schäfer, M. S. (2019). Between guilt and obligation: Debating the responsibility for climate change and climate politics in the media. *Environmental Communication, 13*(6), 723–739.

Prentice, D. A., & Miller, D. T. (1993). Pluralistic ignorance and alcohol use on campus: Some consequences of misperceiving the social norm. *Journal of Personality and Social Psychology, 64*(2), 243.

Raghu, S. J., & Rodrigues, L. L. (2021). Developing and validating an instrument of antecedents of solid waste management behaviour using mixed methods procedure. *Cogent Psychology, 8*(1), 1886628.

Rathi, A. (2023). *Climate capitalism: Winning the race to zero emissions and solving the crisis of our age.* Greystone Books Ltd.

Ressa, M. (2022). *How to stand up to a dictator: The fight for our future.* WH Allen.

Reynolds, R. (1994). *Super heroes: A modern mythology.* Univ.

Ritchie, H. (2021). Stop telling kids they'll die from climate change. *Wired*, 1 November. https://www.wired.co.uk/article/climate-crisis-doom

Ritchie, H. (2022). *Young people feel like they have no future due to climate change; we need to change the narrative.* Sustainability by numbers Substack blog, December 18. https://hannahritchie.substack.com/p/young-climate-anxiety?utm_medium=web

Robertson, C. (2022). *How people access and think about climate change news.* 2022 Digital News Report, Reuters Institute at Oxford University. https://reutersinstitute.politics.ox.ac.uk/digital-news-report/2022/how-people-access-and-think-about-climate-change-news

Rogers, R. W. (1983). Cognitive and psychological processes in fear appeals and attitude change: A revised theory of protection motivation. In J. T. Cacioppo & R. E. Petty (Eds.), *Social psychophysiology: A sourcebook* (pp. 153–177). Guilford.

Rubin, A. (2023). A climate warning from the cradle of civilisation. *The New York Times*, 29 July. https://www.nytimes.com/2023/07/29/world/middleeast/iraq-water-crisis-desertification.html

Sharot, T. (2011). The optimism bias. *Current Biology, 21*(23), 941–945.

Shaw, A. (2020). Promoting social change-assessing how twitter was used to reduce drunk driving behaviours over new year's eve. *Journal of Promotion Management, 27*(3), 441–463.
Simpson, D. (1992). Raymond Williams: Feeling for Structures, Voicing" History". *Social Text, 30*, 9–26.
Tangney, J. P., & Dearing, R. L. (2003). *Shame and guilt.* Guilford Press.
Whitmarsh, L. (2009). Behavioural responses to climate change: Asymmetry of intentions and impacts. *Journal of Environmental Psychology, 29*(1), 13–23.
Williams, D. M. (2010). Outcome expectancy and self-efficacy: Theoretical implications of an unresolved contradiction. *Personality and Social Psychology Review, 14*(4), 417–425.
Williams, D. M., & Rhodes, R. E. (2016). The confounded self-efficacy construct: Conceptual analysis and recommendations for future research. *Health Psychology Review, 10*(2), 113–128.
Williams, R. (1983). *Culture and society, 1780–1950.* Columbia University Press.
Yadav, R. P., & Kobayashi, M. (2015). A systematic review: Effectiveness of mass media campaigns for reducing alcohol-impaired driving and alcohol-related crashes. *BMC Public Health, 15*, 1–17.
Yanovitzky, I. (2002). Effect of news coverage on the prevalence of drunk-driving behavior: Evidence from a longitudinal study. *Journal of Studies on Alcohol, 63*(3), 342–351.

CHAPTER 8

Conclusion

Climate change is a mammoth challenge—for the scientists, industries, institutions, governments, civil society as much as for the journalists and other media professionals covering it. This book offers an interdisciplinary analysis of the concepts from social psychology, sociology, cultural and political studies, among others, to identify the types of storytelling, framing devices and approaches to climate coverage that can maintain the interest of the audience in the climate story. This book provides the conceptual as well as practical solutions as to how to mitigate the effects of eco-anxiety and build the eco-empowerment of the global audiences.

It is crucial that the media generate a solid comprehension of 'common sense' on climate change. There is a massive power in the results of public opinion surveys and the stories that reinforce the idea that the majority of people in many Western countries are very much aware of climate change and are interested in acting upon it and mitigating it. Yet this powerful consensus is hardly seen in the media—this fundamental piece of knowledge is taken for granted, while more awareness of the thoughts and feelings of the close and distant others is crucial in our globalised, overwhelming media ecologies.

There is a variation between climate reporting in the Global North and Global South. As imperfect as this distinction is—it essentially puts a great variety of countries with different political and socioeconomic systems in either of the two boxes—it nonetheless allows to distinguish a range

© The Author(s), under exclusive license to Springer Nature Switzerland AG 2025
A. Denisova, *Effective Climate Communication*,
https://doi.org/10.1007/978-3-031-67340-5_8

of patterns, albeit schematically. There is a heavy reliance on Western media agenda among the countries with less developed or resource-poor media systems. The newsrooms in many developing countries lack scientific literacy to enable confident climate attribution, and they do not have the resources to send more reporters to examine the effects of climate change on their local areas, communities and businesses. They rarely possess the financial means to invest in lavish, beautifully produced documentary, interactive long features, or similar powerful acts of storytelling. The analysis demonstrates an alarming issue of what I call 'delocalisation' of climate change coverage, which is bitterly ironic given that the harshest effects of climate change are most felt in those countries that do not have enough resources to write local stories about them. A further constraint of coverage lies in the reliance of Global South media outlets on the news wires from large elite gatherings (such as COP) or the release of scientific reports such as that of the IPCC. While these reports and events undeniably speak to the needs and worries of Global South countries, they are not contrasted enough with the climate agenda generated from within the affected countries. However, the positive achievements of the climate reporting in the developing countries are also pronounced and vivid—the reporters demonstrate excellent knowledge of their audience and, whenever feasible, offer a varied range of media artefacts that cater to the readers with varying literacy skills and show experimentation with multimedia formats including audio drama and podcasts.

The climate coverage in the so-called Global North is currently at a fragile balance. It demonstrates the traits of consensus—as most citizens and media outlets agree about the anthropogenic nature of climate change and the need to act upon it. The perspectives on the types of action vary—in some countries, such as Italy, the research shows heavy reliance on the political elites and the expectations that they know what they are doing and will fix the crisis; the same level of citizen passivity is observed in other European nations, too. There is a great variety of media stories that tap into political, societal, economic and lifestyle developments surrounding sustainable living. There is no shortage of the exciting, engaging, accessible storytelling that recommends the tweaks to the everyday lives of Western citizens in their pursuit of climate change mitigation. Most of these advices are published by the broadsheets and quality publications, but even tabloids and more popular outlets serve some climate-related storytelling and make it relatable to their audience.

8 CONCLUSION

The caveat of climate communication in the West lies in the term 'greenwishing' that I proposed in this book. It indicates a low-level intervention in unsustainable lifestyle, a wishful thinking approach that nominates an inconsequential act—e.g. choosing a digital menu in a restaurant over a paper-based one or cooking with the banana peel—as a legitimate move against climate change. This type of trivialising, under-researched reporting risks lulling readers into complacency and adding more confusion about the reasonable and effective ways of addressing the crisis.

The discussion of greenwashing—a common term that denominates a sadly common activity—is presented in the middle of the book. It points to the growing field of misinformation that creates a veneer of virtue around the questionable business models and brand identities and presents the environmentally damaging companies and institutions as sustainable and contributing to the climate action. The tactics and communication strategies of greenwashing are becoming more and more varied and sophisticated—and while the regulators are trying to control and address the lies and exaggerations, they always seem to be one step behind.

Some of the more positive insights of this book lie in the field of self-efficacy studies. It is highly recommended to protect and nurture the sense of efficacy in the populations that need to address climate change. Bandura's term refers to the feeling of being capable of addressing the issues at hand by the application of necessary efforts to it. It is paramount to cultivate the sense of efficacy in the audience—it is encouraged to keep praise of the small acts of climate action, even if they comprise consumer choices and workplace decisions. Building on these small steps is the healthy strategy of maintaining the audience's interest in climate change and larger developments in its mitigation. The issue is so mammoth, it is necessary to break it down into accessible chunks of individual and community actions. An alternative—a withdrawal from climate news or neglect of the audience's needs and capacities—will help no one.

Among the techniques that lead from eco-anxiety to eco-empowerment are the focus on solutions or recommended paths to take to minimise the effects of the crisis; the focus on the stories of empowerment and solutions; appealing to the values of universalism and conservatism, not the feeling of guilt and shame, when engaging the readers with the topics; and providing powerful acts of story-telling—strong artefacts that rely on powerful reporting and enchanting execution. These should steer away from apocalyptic undertones and doomist framing—as the narrative of doom and gloom is reducing the sense of agency and opens the doors for fake prophets and populism.

Overall, this book is a journey of discovery—of what the humankind has achieved in terms of science of storytelling, understanding of identity components and action motivators, psychological triggers and inhibitors of action, all applied to the urgent need of adjusting climate change communication to the challenges of the abundant, notifications-heavy media system of our days. It is hoped that this book serves as a source of pragmatic optimism to the academics, media practitioners, stakeholders and any person who feels overwhelmed and disempowered by some of the climate narratives. It illuminates the avenues for efficacy, belief in the power of informed agency and democratic societies where citizens feel that they matter—and use the power they hold to bring a positive change.

Index

A
algorithm, 9, 18, 19, 131, 135
anger, 14, 17–19, 132, 159, 175, 182
anxiety, 2, 3, 7, 13, 14, 17–19, 26–28, 31, 32, 98, 99, 108, 132, 141, 157, 159–161, 164, 175
apocalyptic storytelling, 7, 163, 168–171
Attenborough, David, 183, 184
awe, 3, 17, 19, 24, 32, 37, 38, 132, 170, 188

B
Bandura, Albert, 7, 162

C
carbon footprint calculator, 5, 87, 89
carbon offset, 3, 5, 36, 90
chronotope, 2
Cialdini, Robert, 125
climate attribution, 29–31, 198
community as a hero, 187
conditional cooperators, 119–121

crisis, 1–4, 8, 9, 13, 14, 16, 21–23, 27, 31, 32, 61, 62, 65, 67, 71, 73, 75, 89, 100, 102, 105, 106, 108, 117, 132–134, 141, 145, 158–161, 163, 164, 166, 171–173, 175, 179, 183, 198, 199
cultural relatability, 24, 38, 40, 41

D
documentary, 4, 42, 142, 183, 198
doomism/doomist, 7, 77, 170–173, 175, 200
drunk-driving campaigns, 7, 178, 179

E
eco-anxiety, 49, 159, 161, 197, 199
eco-empowerment, 161, 197, 199
The Economist, 20, 188
emotional offsets, 23, 33
Entman, Robert, 48, 76, 169, 170
Extinction Rebellion, 143, 160, 171

F

Festinger, Leon, 109
framing, 2, 6, 21, 66, 67, 76, 79, 88, 101–103, 128, 129, 133, 146, 169, 170, 172, 197, 200
free-riders, 6, 116, 120, 121

G

Greenpeace, 145–147, 149
greenwashing, 3, 5, 7, 24, 35, 36, 50, 84–92, 199
greenwishing, 7, 24, 92, 93, 199
the Guardian, 1, 30, 50, 71
guilt, 8, 22, 91, 140, 159, 175, 180–182, 199

H

Hayhoe, Katharine, 6, 116, 142
hegemony, 22, 38, 39
hero storytelling, 7, 182, 187

I

identity threat, 21, 142–144
information obesity, 97
information overload, 7, 26, 27
innovation, 3, 16, 21, 23, 24, 31, 34–36, 66, 67, 83, 87
IPCC, 5, 31, 33, 48, 50, 65, 66, 78, 146, 172, 198

K

Kahneman, Daniel, 101, 104, 106, 135

L

Leventhal, Howard, 157
lifestyle/consumer journalism, 43, 44, 46, 92

literacy, 4, 6, 18, 29, 64, 78, 87, 91, 97, 105, 184, 198
loss and damage fund, 78

M

mental health, 14, 103, 141, 175, 183
Monomyth, 183, 187

N

narrative, 1–3, 6–8, 10, 14, 22, 24, 28, 45, 63, 66, 74, 100, 109, 116, 123, 124, 126, 132, 140–143, 162, 171, 173, 181–187, 200
news values, 3, 4, 7, 14–20, 23, 25, 38–40, 67, 77, 86, 160
New York magazine, 169, 170
The New York Times, 26, 30, 50, 69, 71–73, 83, 87, 88, 132, 133, 168–170, 174, 175

O

1.5C, 172, 173
Oreskes, Naomi, 5, 87, 88, 91
Othering, 2, 6, 41

P

pluralistic ignorance, 121, 122
polar bear, 145, 147, 149
public opinion surveys, 120, 124, 197

R

radio drama, 64, 69, 79
Rolling Stone, 189

S

Seaspiracy, 189
second-hand shopping, 113, 114

self-efficacy, 7, 45, 100, 161–168, 175, 177, 178, 199
shame, 8, 22, 180–182, 199
social comparison, 6, 109, 126
social identity theory, 6, 138–140, 144
social proof, 6, 7, 109, 113, 124–126, 130, 131, 167, 174
'the structure of feeling', 164

T
Tajfel, Henri, 138–140
Thunberg, Greta, 160, 183–187

V
values-beliefs-norms (VBN) theory, 115
'view from the moon', 8, 19, 20, 188–190
viral cultures, 5

W
Weick, Karl, 134–136
Williams, Raymond, 164

GPSR Compliance
The European Union's (EU) General Product Safety Regulation (GPSR) is a set of rules that requires consumer products to be safe and our obligations to ensure this.

If you have any concerns about our products, you can contact us on

ProductSafety@springernature.com

In case Publisher is established outside the EU, the EU authorized representative is:

Springer Nature Customer Service Center GmbH
Europaplatz 3
69115 Heidelberg, Germany

www.ingramcontent.com/pod-product-compliance
Ingram Content Group UK Ltd.
Pitfield, Milton Keynes, MK11 3LW, UK
UKHW031557170425
457567UK00009B/43